Comic Character

TIMEPIECES

SEVEN DECADES OF MEMORIES

Hy Brown with Nancy Thomas

Schiffer Publishing Ltd

1469 Morstein Road, West Chester, Pennsylvania 19380

Dedication

I sit here on my 49th birthday promising to start and complete this book as a 50th birthday present to myself. In that endeavor, I hereby dedicate this book to . . .

CAROL CHASE, whose pricing structure forced me into starting all this research;

HOWARD BRENNER, who, by spending all those hours helping me, kept my hobby from ending in its embryonic stage;

GEORGE NEWCOMB, MAGGIE KENYON, DAVE MYCKO, and JEFF COHEN who put up with my voluminous, illegible, handwritten want-lists, and without whom I would have been unable to acquire the majority of my collection;

KEN JACOBS, RON PINKERTON, LINDA & JERRY HOLLEMAN, who not only put up with my lists, but also with me;

DEAN TAYLOR, who taught me the real meaning of the word "mint";

STEVE STEGMAN, who answers the phone the greatest;

SANDY KESSLER, who finally, finally made the deal that got me my two best pieces;

ROBERT LESSER, who started it all;

EHRHARDT'S—all of them—for all their help;

DARREN of Cartoon Junction for finding me the "new stuff";

NANCY PLUMMER who really knows how these things work;

ED FABER who talked me into doing this book;

DARYA KLEIN who pitched in at the end and helped type the captions, a chore unto itself.

IRA SELWIN, my nephew, who, if he continues trading me his best pieces, might someday get the whole collection;

STEWART UNGER, who, after convincing me to attempt this book, supported me in every way a person can and also taught me the meaning of true friendship;

BRIAN, JASON, and MELISSA, my children, who had to live through this second childhood with me;

NANCY, my wife, who gave me back life and then allowed me to use part of it to be a little boy again.

I thank you all. I like you all, but I love you, Brian, Jason, Missy, and Nance.

Published by Schiffer Publishing, Ltd.
1469 Morstein Road
West Chester, Pennsylvania 19380
Please write for a free catalog.
This book may be purchased from the publisher.
Please include $2.95 postage.
Try your bookstore first.

Acknowledgements

Within any hobby, there are and always will be people who have acquired collections that are so significant that not acknowledging them and mentioning them would be an injustice to the new collector. I believe that the collection acquired by Mr. Robert Lesser and sold in the mid-1980s, is such a collection. It is fully showcased in his book, *Celebration of Comic Character Art and Memorabilia*, published by Hawthorne in 1975.

Howard Brenner's collection of over 400 pieces, all of which were mint-in-the-box, also qualifies. Mr. Brenner's collection is included in his book, *Comic Character Clocks and Watches*, published by Books Americana in 1987.

Special thanks are hereby given to my son Brian who, by editing this book, was able to continually criticize his father without fear of retaliation, a process I am sure he found thoroughly enjoyable. On the other hand, I am not sure which I found more enjoyable—the realization that my son actually learned something in college, or that he learned it better than I.

Special thanks are hereby given to my wife Nancy who, by helping me with this book, realized that it is actually possible for an engineer to write a sentence without punctuation or paragraphs that can run on for fifty pages, something I am sure she found totally unenjoyable. I, on the other hand, finally realized the value of a liberal arts education; periods, commas, and paragraphs do help. I have no problem accepting the fact that my wife learned it better than I.

Preface

Some people can look at an object and decide if they like it or not, but I am not able to do that. I must first ask others what they think, then try to figure out what I think, and finally force myself into making a decision. Because I am required to make numerous decisions every day, I try to minimize the decisions I need to make with reference to my hobbies. In order to accomplish this, I start by making a list. After completing the list, the only decision I need to make is, "do I want to start collecting these items?" After this, no decisions are necessary because I simply collect everything on the list. I have traded research for decision-making. This book is being written for those people like myself who cannot decide what they like and do not want to make any decisions that they do not have to make. For those of you who do not have this problem, I hope you enjoy it anyway. You do not have to study any of my lists!

Contents

Introduction

In 1933, the Ingersoll-Waterbury Company produced a round Mickey Mouse wristwatch that is considered to be the first comic character wristwatch. The Ingersoll-Waterbury Company continued to produce this style of watch through 1937. In 1938, sensing that the public tired of this model, Ingersoll introduced a rectangular-shaped Mickey watch. There were, however, round watches in the company's inventories that had already been made, but had not been sold. These were still available through mid-1940.

Make a watch until people stop liking it, and then change it! This approach became the marketing theory of watches through 1972. After 1972, the theory changed. Instead of making one watch and waiting until it faded from popularity, someone decided to make twenty different models of each character. The theory was that because of more choices, more people would buy more watches!

For the collector, this means that, because there is a definable, limited number of watches produced during the "golden years" 1933-1972, it is not necessary to choose the best or the most collectible. Instead one can make a list of all 375 watches that were made during this period. (Remember! The beauty of the hobby is that previously unknown models are sometimes discovered!) Obviously, this is not practical with respect to the more than 5,000 watches made since 1972.

Comic Character Timepieces is divided into nine chapters. Except for the chapter on political and personality watches (Chapters 8 and 9), all items will be listed alphabetically by the first name of the character. The politicians and personalities are listed alphabetically by their last names. For example, Wile E. Coyote is listed under "W" and Foghorn Leghorn is listed under "F", while Richard Nixon is listed under "N". Also included in each chapter are pocket watches and clocks from that same period.

At the end of each picture caption, you will find a letter: "S", "M", or "L". This is to designate the size of the watch face. "S" is a diameter of 25-28 mm; "M" is 29-32 mm; and, "L" is 33 mm or greater.

Throughout the chapters there are advertisements from the period and pictures of the original boxes or packaging. Also included are anecdotal stories, such as why, more often than not, the Red Rider (1951) dial is superimposed on top of the Bugs Bunny dial.

Prior to 1933, manufacturers produced clocks and pocket watches that were forerunners to the comic character watches and timepieces that began to be mass-produced that year. Chapter 1 will review the pocket watches and clocks that were made during this earlier period.

Chapter 2 reflects the watch-making years between 1933 and 1939. In 1933, Ingersoll Watch had approximately 300 people working for them. They then began producing Mickey Mouse, Donald Duck, and Big Bad Wolf watches. Within two years employment had increased to 3,000 and over 2,500,000 watches had been produced. In addition to Ingersoll, other watch companies entered the market. Ingraham started making Buck Rogers and Betty Boop watches. New Haven Watch Company joined the fray with watches such as the Lone Ranger, Dick Tracy, Orphan Annie, and Smitty. The popularity of comic watches continued to grow until it was abruptly halted by the Second World War. Watch company plants were converted to military use, and production did not begin again until 1946.

Chapter 3 describes the years between 1946 and 1958. When production resumed, in 1946, watch companies not only reissued the watches that were being made before the war in smaller sizes, but they also added new characters such as Captain Marvel, Roy Rogers, Bugs Bunny, Joe Palooka, Howdy Doody, and Woody Woodpecker. In 1957, Ingersoll (then operating as U.S. Time) personally presented its 25,000,000th watch to Walt Disney. However, by the end of 1958, it was widely apparent that the demand for comic character timepieces was at an end. U.S. Time ceased production in character watches, and it was not to begin again for ten years.

Chapter 4 covers the years between 1958 and 1972. Between 1958 and 1968, Bradley Watch Company produced a minimum number of character watches, such as Popeye and Quick Draw McGraw, but there were no Disney watches made in this country during the ten-year period. The Japanese, however, entered the market and filled the void created by U.S. manufacturers.

Things changed in 1968, when a nostalgia craze swept the United States. Timex (U.S. Time's new name) began producing Mickey Mouse and other Disney characters again. In 1971, Jay Ward of Rocky and Bullwinkle fame, produced a series of 16 watches. Also, Sheraton Corporation, as an advertising inducement, joined the fray with a series of watches portraying Hanna-Barbera characters. The Sheffield Watch Company began their own series of watches that included characters such as Felix the Cat and Popeye. The years 1968-72 were interesting times. Many new companies entered the market with new models and different types of watches, such as models that were battery-operated and the first comic character watches to have a 17-jewel movement. The period came to an end in 1972, when the Bradley Watch Company took over the production of all Disney character watches and began the era of each individual character being made in many different designs each year.

Chapter 5 includes the years 1973 through 1985, often referred to as "the Bradley Years." Though other companies were producing watches, Bradley dominated the market, designing approximately 2,000 different models. This chapter in no way attempts to show or list them all, but only to give you a flavor of the vast number of watches that were made and are available to collectors.

Chapter 6 highlights the comic character watches from 1985 to present. Seiko was awarded the exclusive contract to produce Disney watches in 1985 (except for the theme parks and Disney Stores) and continued a marketing strategy identical to the one that Bradley had successfully implemented. But in addition, Seiko added its own special designs, such as combining characters which resulted in the first Mickey and Donald, Mickey and Pluto, etc. The exclusive Seiko contract was lost in 1990, and many companies are now authorized to produce Disney watches. During these years, new companies have entered the market, including Armitron, Hope Industries, and Innovative Time. Old characters were reintroduced, and new characters, such as Mutant Ninja Turtles, Tiny Toons, Tale Spin, and Ghostbusters, were produced for the first time. Fantasma introduced 3-D hologram watches, animated disc, and musical computer disc watches. This chapter, like Chapter 5, will only give you a flavor of what is available.

Chapter 7 describes a different but related type of watch. Advertising watches started with the advent of Buster Brown, and continued with such characters as Charlie the Tuna, Bud Man, and Tony the Tiger. Familiar corporate symbols like the Ritz Cracker, Coca-Cola, 7-Up, and Heinz Ketchup were placed on the face of a watch. Each week, advertising agencies would invent new characters or reuse old ones to sell their goods. In an effort to stimulate sales, the companies would sometimes place this new advertising on watch faces and make them available to the public.

Political watches comprise Chapter 8. These watches include the infamous Nixon Moving Eyes, Jimmy Carter with the body of a peanut, Ronald Reagan riding an elephant, and my favorite, Sam Erwin where each number on the dial is the name of one of the Watergate 12. Political watches even attempt to make a statement such as the Teddy Kennedy waterproof watch, the Gerald Ford on skis wearing a football helmet, and Lester Maddox with a bat in one hand and a drumstick in another. It is interesting to note how these watches reflect a piece of our history.

Through the years, some of our favorite comedians, actors, actresses, and personalities have adorned the faces of watches. Chapter 9 is a comprehensive list of these types of watches. It includes such notables as W. C. Fields, Buster Keaton, Abbot and Costello, Laurel and Hardy, Jerry Lewis, Elvis Presley, Lucille Ball, Marilyn Monroe, and Whoopie Goldberg. For all we know, many of these celebrities have had their own personalized watch and have enjoyed wearing it.

Included at the end of each chapter is a section of lists in an effort to help the collector start his or her collection. Each collector has a slightly different approach to their collection. There are those who wish to collect only those watches made in the year of their birth. Some will want to collect only Mickey Mouse watches. Other collectors may want only those watches produced prior to 1972. Western fans will try to collect only cowboy watches. And, then there are those of us crazy individuals who try to collect them all.

Collecting comic character timepieces can be accomplished in one of two ways. One group collects them "mint-in-the-box." To these collectors, the Hopalong Cassidy saddle, the Davy Crockett powderhorn, and the Cinderella glass slipper are an integral part of the watch, and are displayed as pieces of art. To the "mint-in-the-box" collector, everything must be as it was originally designed in order to be considered authentic.

Another group of collectors wears their watches, a different one everyday. In order to accomplish this, one of the first things one must do is change the band. Companies originally produced the watches for children. To the wearer of comic watches, the face, the hands, the case, and the movement must be the original, but it is not necessary that the band or the crystal be authentic.

I have mentioned that there are approximately 375 different watches between 1933 and 1972, and this book will list all of these watches and will attempt to show a picture of each one. However, what makes a watch different is a matter of judgment, not of fact. Some watches come in both a gold-toned and silver-toned case. Does this difference mean that the watches are two distinct models? The answer...sometimes.

The 1938 gold Mickey, the 1947 gold Mickey, and the 1947 gold Donald Duck were marketed as deluxe models and sold for $1.00 more than their silver companions. Consequently, collectors consider the watches to be different models. In contrast, the mid-1950 Roy Rogers gold and silver watches were marketed in identical ways and, therefore, are not considered to be different. The Dale Evans watches came with coordinated cases and bands, that is, pink band/pink case, blue band/blue case, etc. Each coordinated combination is considered to be a different watch even though the marketing is identical to the Roy Rogers watch.

Should a case variation make a watch different? Again, the answer is sometimes. The difference between the 1934 Dick Tracy case and the 1935 Dick Tracy case (which was an indentation around the crown), is not significant enough for me to consider the watches to be different. However, the differences in the fluting between the 1948 and 1949 Disney watches is significant enough for these watches to be considered different.

Except to the mint-in-the-box collector, watchbands are usually not considered to be an integral part of the watch. However, there are exceptions and they include the aforementioned Dale Evans watch, the 1933 Mickey, the 1934 Three Little Pigs/Big Bad Wolf, and the 1934 Tom Mix.

This book attempts to list all of the different watches between 1933 through 1972, but for the years 1973-1985 I have had to take a different approach. For this period, I am making a guess about which watches the collecting public will someday consider to be significant. There are some watches from this period that have already reached that level: the Super Hero watches of the late 1970s produced by Timex and Dabs, the 50th anniversary Mickey Mouse watches, etc. Only the future can tell which watches of this period will be considered the BEST.

From 1985 to the present, I have attempted to show what types of watches are available. Obviously someday the 60th anniversary Mickey Mouse will be considered collectible. The Armitron series of Bugs Bunny, Porky Pig, Elmer Fudd, and the twelve others are certainly desirable. Without question Roger Rabbit and Jessica Rabbit will be considered classics. The Mutant Ninja Turtles, Duck Tales, Tiny Toons, and other plastic molded watches being produced by Hope Industries and Innovative Time might someday be considered collectible. We even have our first verifiable collector's watch from this period—the limited edition gold, backwards Goofy watch produced by Pedre which sold for $75 has already changed hands for as much as $500.

While this book includes a price guide, the personal value is only in the eyes of the buyer and the seller. This book is written to enlighten, illuminate, and to induce others to enjoy the comic watch hobby. Comic character timepieces can be found at watch shows, toy shows, swap meets, antique and collectible shows, and in everyone's attic. Enjoy and good hunting!

FINAL THOUGHTS

It is now eight months and thirteen days since I began writing this book, and it appears that my desire to have it published by my 50th birthday is within reach. During my research, many people were kind enough to share with me their knowledge and thoughts. An overwhelming majority of them asked the same five questions, which I would like to answer here.

1. How many watches do you have in your collection? *As of this date, 1396.*

2. Have you found any watches that you did not know existed? *Yes. During the past eight months I have added five new watches for the period 1933-1972.*

3. Do you have in your collection all the watches from this period? *No, I am missing 23.*

4. Do you plan to continue your research and update your list? *Yes! In order to accomplish this, however, I would appreciate anyone's input. If you would like to ask questions or send me pictures of watches not in this book, please feel free to do so by mailing them to the address listed below.*

5 How will we be able to obtain the latest information? *If there is a demand for this knowledge, I would be more than happy to publish a newsletter with this information. If there is really a large demand, an updated version of the book will probably be published at a future date.*

For those who have hobbies with knowledge that you feel would benefit others, writing a book can be a pleasurable experience, IF you are fortunate to have a publisher like Schiffer. I am proud to say that both Nancy and Peter Schiffer have become my friends. Their support, understanding, and knowledge was not only invaluable but also inspiring. In addition, their courtesy and hospitality gave me a new meaning to the words "nice people."

My editor is Douglas Congdon-Martin: he who held my hand; he who kept telling me to go with my gut; he who took all the photographs; and he who made doing this an almost painless effort. He makes me proud to call him my friend also.

A book of this nature cannot be accomplished by only one person. It is like building a table that needs four legs. One leg is the author; the second leg is the publisher, Peter Schiffer; the third leg is the editor, Douglas Congdon-Martin; and the last leg is Nancy Thomas, my wife. The four re-editings that Nancy did, turned the gibberish into English so that Doug was able to edit and complete this book in our lifetime. Nancy's patience and knowledge was an integral part of the completion of this book.

Would I do it again? At the drop of a hat, but only if the other three legs of the table agree to help.

Hy Brown February 29, 1992
22343 Gilmore St.
Canoga Park, California 91303

Chapter One
Early Novelty Timepieces:
Prior to 1933

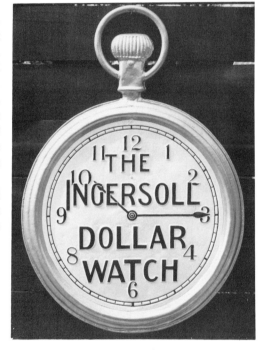

In 1880 two brothers, Robert and Charles Ingersoll, were making watches and clocks that sold for $1.00. In the same year, Junghans produced a clock titled "Man on a Trapeze." It pictured a man in simulated flight through the air. The Ingersoll's, liking the idea, changed the concept somewhat and issued the first commemorative watch for the Chicago World's Fair in 1893. Junghans continued with their type of animated clock: Lady with a Fan, Lady with a Mandolin, and a Negro banjo player. Ingersoll continued with the 1901 Pan Am Expo watch, and the 1904 St. Louis World's Fair. In 1908 Buster Brown appeared on a pocket watch to advertise shoes. Finally, in the early 1930s, Lux issued a series of clocks such as the organ grinder, and the Negro shoeshine boy.

I am often asked if someone woke up in 1933 and all of a sudden decided to make a Mickey Mouse watch. Obviously not. As you can see, prior to the brilliant idea of mass producing a Mickey Mouse watch, there were political, promotional, and advertising watches.

The Ingersoll Dollar Watch, c. 1880.

Advertising pocket watch, c. 1930. *Courtesy of Roy Ehrhardt.*

Buster Brown. An early pocketwatch that predated the advent of character watches. Size M. *Courtesy of Jack Feldman.*

NOTICE: This Display Case of which Ingersoll Watch Co., Inc., is the owner and lessor, has been leased **solely for use in displaying INGERSOLL WATCHES** and upon the further express condition that the use of it at any time for any other purpose shall immediately terminate the lease and thereupon said owner and lessor shall immediately become entitled to take possession of and remove the case from any premises upon or in which it may then be without incurring any liability for trespass or for the return of any part of the rent theretofore paid for its use.

INGERSOLL WATCH CO., Inc.

Advertisement for the Ingersoll pocket watches in the 1930s, just prior to the advent of the Comic Character Timepieces.

Advertising pocket watch, c. 1930. *Courtesy of Roy Ehrhardt.*

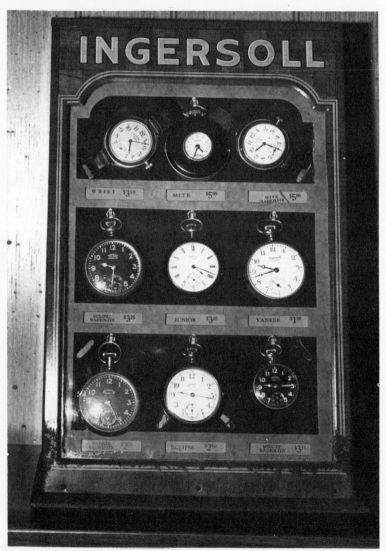

Ingersoll watches, c. 130. *Courtesy of Roy Ehrhardt.*

Advertising pocket watch, c. 1930. *Courtesy of Roy Ehrhardt.*

The beginning—animated character clocks from the nineteenth century: Negro Mammy With Rolling Eyes, Bixby's Best Shoe Polish, and Man on the Trapeze.

Animated clocks produced by Ingrahams, 1880-1895.

Lux animated clocks from the early nineteen thirties: Happy Days clock celebrating the repeal of Prohibition, the Organ Grinder and the Monkey, and the Negro Shoeshine Boy.

Animated clocks produced by Lux, the forerunners to comic character timepieces, 1930.

Chapter Two
The Prewar Years:
1933-1939

Big Bad Wolf, 1934, Ingersoll. The eye blinks and the back of the watch is embossed with the wolf. The watch fob is as collectible as the watch itself. *Courtesy of Jeff Cohen.*

Betty Boop, 1934, made by Ingraham. One of only four known to exist mint-in-box. The back of the watch is embossed with Betty Boop. *Courtesy of Jeff Cohen.*

The year is 1933. You are at the Chicago World's Fair, and you walk past various booths. Inside one booth, men are assembling and selling something called a "Mickey Mouse watch". The watch face is a piece of white paper with a Mickey Mouse image made into a circle. Between Mickey's legs is a small disc with three Mickeys revolving. There is a choice of two bands: one is a metal link with the first link on each side of the dial having embossed Mickeys and the other is a leather band with metal Mickey appliques attached to the band. The cost is $3.95. By the end of the year 1933, 900,000 people had purchased this watch. The comic character watch market had begun. Within only two years, 2.5 million Mickeys would be produced.

What preceded this event? Ingersoll Watch Company had sold 50 million $1.00 pocket watches by the end of World War I. However, hard times hit the country, and by 1922 Waterbury purchased Ingersoll and the new company was called Ingersoll-Waterbury. By 1933, employment at the watch company had decreased to 300 people. Wristwatches were just too expensive for people to purchase. Ingersoll had a large supply of World War I pin lever movements. The company believed that by using this type of cheap movement, an affordable wristwatch could be produced. The idea of putting Mickey Mouse on it added something different to the watch. By the end of 1933, employment at the company had ballooned to 3,000 people. Introduction of the watch at Macy's in New York led to the sale of 11,000 in one day. Grandparents bought the watch for their grandchildren as graduation presents, and grown men bought the watch because they could afford it.

Big Bad Wolf, made by Ingersoll, 1934. The metal link band with the **Big Bad Wolf** on one side and the Three Little Pigs on the other side makes this watch even more desirable. Size M.

Three Little Pigs, advertisement, 1934

Mickey #1, as the above watch is called, came in a red, rectangular box with all of Mickey's friends on the cover-—Minnie Mouse, Clarabelle the Cow, Horace Horsecollar, Pluto, and Mickey himself. In 1934 two changes were made to the watch. The packaging was changed to a blue, flat rectangular box with only Mickey on it, and, to stop counterfeiting, the words "Made in the U.S.A." were added below the number 9. This watch was mainly sold through the Sears catalog with the listed price of $2.95, but it could be found on sale for $2.69. While production of this watch stopped in 1938, the watch continued to be advertised and sold through 1940 by utilizing previously made components.

In England during the 1930s, it was not chic to wear a wristwatch. However, Mickey was popular, and so in addition to producing pocket watches, the British produced their version of the Mickey #1. There are, however, substantial differences. The watch was much smaller in size and was definitely meant to be worn by a child. Mickey was made to look more rat-like and less cutesy. The British could not understand why Mickey's hands were outlined in yellow. Therefore, the British model has orange hands and shows all his fingers. In addition the British felt they wanted Mickey to look more masculine and made one final change. The little red dots on Mickey's chin, on the pocket watch and the wristwatch, are to simulate a three-day-growth.

In 1934, because of Mickey's success, Disney decided to issue the Three Little Pigs watch. It is a round, red-faced watch with a large black wolf prominently placed in the center. The wolf is staring at the three little pigs

Buck Rogers, 1935, Ingraham. Must have bolt action hands and embossed back of one-eyed monster. *Courtesy of Roy Ehrhardt.*

Dan Dare, 1953, Ingersoll. Popular English comic strip hero. This watch is double-animated: Dan's hand goes up and down, and the rocket ship second hand rotates. *Courtesy of Roy Ehrhardt.*

Dick Tracy, New Haven, 1935. The watches of this period, in this shape, were 25% larger than those made after WWII.

Donald Duck, 1935. *Courtesy of Jeff Cohen.*

from the top of the watch. The watch's original band is metal with the first link on one side showing the three pigs and the other side's first link with the wolf. Disney did not want the watch called "The Big Bad Wolf" despite the prominence of the wolf. They wanted it called The Three Little Pigs because in Disney's mind, the wolf was controversial and scary. However, in response to minimal sales, Sears began calling it the Big Bad Wolf watch, and by that name it is known today.

The New Haven Watch Company joined the comic character watch business in 1934. They created Dick Tracy, Orphan Annie, and Smitty, all in the same large rectangular case. While Mickey #1 is referred to as the first comic character wristwatch, there is known to exist an ad for the Orphan Annie watch in a magazine dated January, 1933. This would obviously mean that this watch was being produced prior to the Mickey #1. However, because of history, and for the same reason that people still believe that Abner Doubleday invented baseball, Mickey is still considered the first watch.

In 1935, New Haven introduced what most collectors consider to be the most attractive of the early watches—Popeye. Not only was Popeye's image on the rectangular-faced watch, but also the members of the Thimble Theatre players. Not to be outdone, Ingersoll introduced the 1935 Donald Duck watch. The watch was round and the same size as the Mickey #1, and between Donald's legs was the same disc that was on the Mickey #1. It was not until July, 1990 that we discovered that this watch came with a leather band with metal Donald appliques. (The story concerning this discovery is told at the end of this chapter.)

The Eberite Watch Company entered the fray with the first personality watch, a rectangular Dizzy Dean made in 1935. The first cowboy watch, Tom Mix, was introduced in 1935 by Ingersoll. It pictures Tom Mix on his horse, Tony, and a Texas longhorn on a rotating disc. It has a metal band displaying one link of Tom and one link of his horse, Tony. The watch is highly desirable not only by the watch collector, but also by the cowboy memorabilia collector. Because of this, the Tom Mix watch is only slightly less rare than the 1933 English Mickey or the 1935 Donald Duck.

Donald Duck, made by Ingersoll, 1935. This band has Mickeys on it, but the original had Donald appliques. Only three of these are known to exist. This watch did not go into production. Size M.

Donald Duck, 1939, Ingersoll. Came with a Mickey decal on the back.

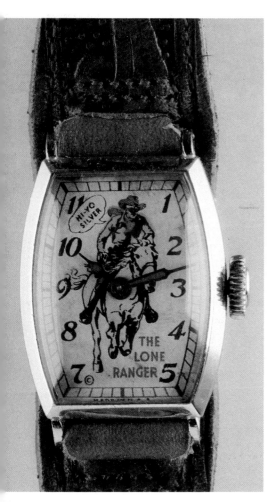

Lone Ranger, made by New Haven, 1939. The watches of this period, in this shape, were 25% larger than those made after WWII.

By 1938, people were tiring of Mickey #1. Ingersoll issued a new model that was rectangular and still retained the disc between Mickey's legs. There was fluting on the metal case next to the numbers 3 and 9. The watch came with a metal or a leather band. With the woman's version of the metal band, one could also purchase charms of the Disney characters. In 1939 the disc was removed and replaced by a blade second hand. The case design was also slightly changed by reconfiguring the fluting. For the first time, a gold-tone model was manufactured and sold for $1.00 more than the silver one. Obviously, the gold-tone is more desirable to the collector. It is this model that was sealed in the capsule that was buried during the 1939-1940 World's Fair.

New Haven finished the 1930s by introducing in 1939 the first of the super heroes and the second cowboy watch, Superman and the Lone Ranger, both in large rectangular cases.

In the event that you were not counting, that makes 15 watches made prior to World War II. Because Mickey interests most people more than the others, I will summarize the pre-war Mickeys: The 1933 English and American, the 1938 Mickey rectangular, and the 1939 Mickey silver and gold.

There are other watches from this period that various collectors claim to have seen, and of which I have seen one. However, because I cannot find a second, I choose to call it a prototype. It is a 1933 Mickey #1, but it is gold-toned and the hands are made of copper. Other collectors claim to have seen a Buck Rogers made in 1938, and a Flash Gordon made in 1939. And of course, anything is possible when collecting comic watches.

Prior to 1939, the favored timepiece was the pocket watch. In 1933 Ingersoll produced two versions of Mickey #1 pocket watches: one with a long stem, and one with a short stem. Both were round and had the famous Mickey disc between his legs. At the same time in England, two versions of the Mickey pocket watch were also produced. One portrayed Mickey as short and fat while the other had a more rounded Mickey. On both, Mickey had his familiar (at least to the British) three-day growth.

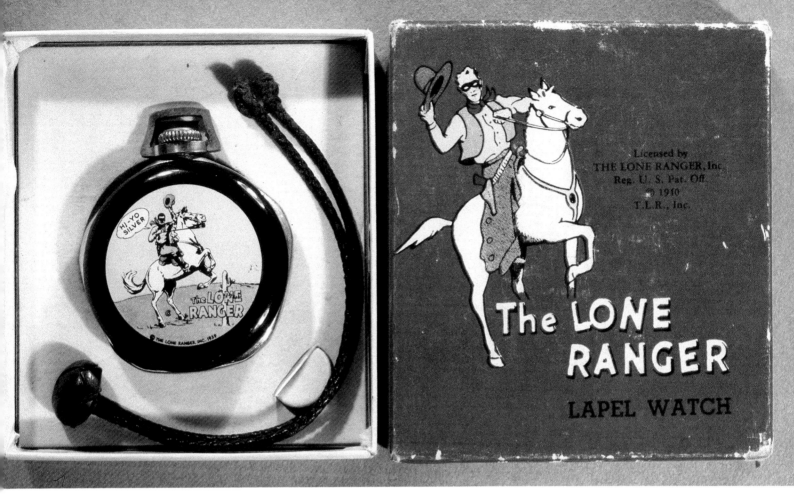

Lone Ranger, 1935, New Haven. The decal is what makes this pocket watch so valuable. *Courtesy of Roy Ehrhardt.*

Mickey Mouse 1933. Ingersoll electric clock featuring a revolving Mickey. There is a decorative strip around the outside of the clock containing all the popular Disney characters. *Courtesy of Jack Feldman.*

Mickey Mouse, 1933. Ingersoll. Commonly referred to as Mickey #1. This is considered the first comic character watch. This watch was made from 1933 through 1937, but only the ones made after 1934 say "Made in the USA" by the number 8. The band is as important as the watch. Size M.

Mickey Mouse, 1933 pocketwatch, considered to be the first one made. *Courtesy of Jack Feldman.*

Mickey Mouse, 1933 Ingersoll Mickey #1. This is the later variation of the 1933 wristwatch. Size M. *Courtesy of Jack Feldman.*

Mickey Mouse, 1938, Ingersoll. Called the Lapel Watch, the decal on the back is what makes many people consider this to be the most attractive of the pre-war pocket watches. *Courtesy of Jeff Cohen.*

Mickey Mouse, advertisement from a Sears catalog, 1935.

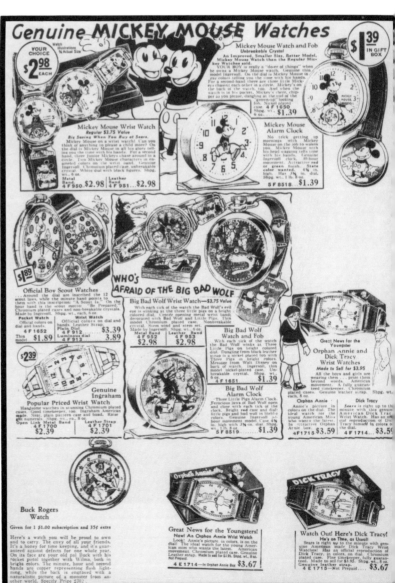

Mickey Mouse, 1933, Ingersoll. This is the model with the short stem. It has Mickey embossed on the back. The watch fob is also very collectible. *Courtesy of Roy Ehrhardt*

Mickey Mouse, 1933. Original display of Mickey #1 watches and the original box. Size M. *Courtesy of Jack Feldman.*

Mickey Mouse, 1933 English Mickey. The English decided to give Mickey a little three-day growth and fingers. Size S.

Mickey Mouse, 1939, in box.

Mickey Mouse, 1939. Ingersoll. This is the '39 Mickey in the '38 case. These watches were actually produced this way in 1939. Size L.

Mickey Mouse, 1938, in box.

Mickey Mouse, 1939 Ingersoll Deluxe model. This gold-tone model was made in a limited edition. Notice the fluting at 3 and 9 and also the blade second hand. These are both as advertised for the 1939 watch. Size L.

In 1934 New Haven produced a Popeye pocket watch and surrounded him with the Thimble Theatre players. Ingersoll joined the fray with a 1934 Tom Mix pocket watch. In 1935, for some unknown reason, New Haven reissued Popeye without the Thimble Theatre players and only showed Popeye's face.

Eberite produced the Dizzy Dean in 1935 and Ingraham Watch Company produced Buck Rogers with bolt action hands and a one-eyed cyclops engraved on the back. Ingraham also produced pocket watch versions of Betty Boop, Rudy Nebb, Skeexix, and Moon Mullins. None of the Ingraham pocket watches were made into wristwatches, a fact that we wristwatch collectors bemoan.

There was a 1936 New Haven Smitty, and in 1937 Ingersoll produced the Big Bad Wolf to match the wristwatch. This pocket watch was unique in that the wolf's eyes blinked, and the back of the watch was embossed with the wolf. In 1938 Ingersoll produced a black Mickey lapel watch that has a Mickey decal on the back. The watch also came with a fob. Most collectors consider this to be the most beautiful of all the Mickeys. New Haven and Ingersoll ended the 1930s by producing the 1939 Lone Ranger and the 1939 Donald Duck. There are those who claim that there are two more in addition to those I have listed. Could there really be a Flash Gordon or a Yellow Kid out there somewhere? I have been keeping count for you—20 known pocket watches.

There were eleven comic character clocks made prior to 1940, seven of which were of Mickey Mouse. Ingersoll made two Mickey clocks in 1933. One was a square electric model with a wagging Mickey head. A second, while square, was not electric and had a revolving disc between his legs. Ingersoll, realizing that they did not make an alarm clock in 1933, corrected this in 1934 with a round, wagging head Mickey. Also in 1934, they made an art deco desk model. England produced a Mickey alarm and a Mickey wind-up, also in 1934.

The Big Bad Wolf clock of 1934 matched both the pocket and the wrist watch, but on the clock the wolf's jaw moves. One of the nicest displays is to see all the Big Bad Wolf timepieces together--wristwatch, pocket watch, and clock.

Mickey Mouse, 1938. Ingersoll kept the circular disk second hand from the '33 Mickey. The fluting at 3 and 9 was only done this way for one year and changed in 1939. This watch came in a deluxe woman's model with a metal band and charms of Mickey and Minnie. Size L.

Mickey Mouse advertisement from Ingersoll, 1936.

Mickey Mouse, advertisement for a new Mickey Mouse alarm clock and other watches. 1935.

Mickey Mouse advertisement, 1935.

Orphan Annie,1935. Made by New Haven. The watches of this period, in this shape, were 25% larger than those made after WWII.

One of everyone's favorites is the 1935 Popeye and his friends. In 1939 Gilbert Watch gave us Charlie McCarthy in a white enameled case with a puppet on a clear celluloid disc. The 1930s ended with the 1939 Donald Duck by Ingersoll. Another Mickey clock is purported to be a wall clock made by Bayard in the mid-1930s. Again, this is one of those that only a few have claimed to see. Does this Mickey really exist, and is it six or seven pre-war Mickey clocks?

In July, 1990, comic character watch collectors across the country were receiving excited phone calls. At the flea markets of Brimfield, Massachusetts, a 1935 Donald Duck wristwatch, mint-in-the-box, had appeared. To everyone's amazement, it had an unheard of variation. Up to this point, there were only two 1935 Donald Duck wristwatches known to exist, one in the Disney Archives in Burbank, California, and the other, which had been part of Mr. Robert Lesser's collection, was displayed in Samuel's Museum of Comic Character Art in St. Louis.

This watch was round and the same size as the Mickey #1. Between Donald's legs was the same rotating disc of three Mickeys that existed on the Mickey #1, and on the leather band were the same Mickey appliques. The story told was that in 1935 they were not sure that Donald would be a salable commodity, so the familiar Mickey disc and the band with Mickeys were included. For the previous 55 years, this was the only way the watch was known to exist. However, the Brimfield Donald, mint-in-the-box, had metal Donalds on the band!!

At approximately the same time that the Donald appeared in Brimfield, the Samuel's Museum sold its watch to a collector in Chicago. In England, the only known English Mickey #1 was auctioned and bought by a dealer in Philadelphia. This same dealer purchased the Brimfield Donald. Until this time, there had never been a sale of a wristwatch out of the box for over $1,000, and there had never been a sale of one in a box for over $2,000. After four and one-half months of negotiations, I acquired both the Lesser Donald Duck and the English Mickey for $4,700, and the Chicago collector acquired the Brimfield Donald for over $6,000. For a private sale, this was astronomical for its time, and it is almost impossible to place a value on these watches if in fact they were to be auctioned. Nonetheless, I just had to have them!!

Popeye, made by New Haven, 1935. One of the large watches made prior to WWII.

Popeye, 1934, New Haven. This pocket watch shows the Thimble Theatre Players, unlike the 1935 model. *Courtesy of Jeff Cohen.*

Popeye, 1935, New Haven. This is the Popeye without his friends. *Courtesy of Roy Ehrhardt.*

Popeye, 1934, New Haven. Considered to be the most attractive of the pre-war clocks. *Courtesy of Roy Ehrhardt.*

Rudy Nebb, 1933, Ingraham. One of the earliest pocket character watches, this one was never made into a wristwatch. *Courtesy of Roy Ehrhardt.*

Smitty, made by New Haven, 1935. The watches of this period, in this shape, were 25% larger than those made after WWII.

Three Little Pigs, 1939. Prototype pocket watch that never went into mass production. *Courtesy of Jeff Cohen.*

Superman, 1939, in box. *Courtesy of Jeff Cohen.*

Superman, 1939, New Haven. The watches of this period, in this shape, were 25% larger than those made after WWII.

Throughout this chapter, you will see examples of watch packaging. To almost all collectors, even to those of us who wear our watches, the boxes are considered to be pieces of art. The 1938 Mickey deluxe, showing him in a top hat, and the **Big Bad Wolf** are both attractive and highly desirable.

One of the greatest pleasures of this hobby is the constant discovery of what was or what was not produced. No sooner do you feel that you know what exists that there is another watch discovered. For instance, even as this book is being written, I am on my way to New York to be shown another copy of the 1933 gold-cased Mickey. If this is true, contact me so that you can change all the numbers that I gave you above by one!

THE PREWAR YEARS: 1933-1939

WRISTWATCHES

NAME	YR	NOTES
Big Bad Wolf	34	red with band
Dick Tracy	34	large size
Donald Duck	35	Mickey disc/leather band w/ Mickeys
Lone Ranger	39	large size
Mickey Mouse	33	English w/beard
	33	gold case & band/copper hands
	33	with metal band
	38	disc/women bracelet
	39	gold/blade
	39	silver/blade
Orphan Annie	34	large size
Popeye	35	with friends
Smitty	34	large
Superman	39	large size
Tom Mix	35	with embossed band

CLOCKS

NAME	YR	NOTES
Big Bad Wolf	34	Ingersoll/jaw opens
Charlie McCarthy	38	Gilbert/mouth opens
Donald Duck	34	Ingersoll
Mickey Mouse	33	Ingersoll electric/square/moving head
	33	Ingersoll/wind-up/square/with disc
	34	art deco/desk model
	34	English/alarm
	34	English/wind-up
	34	Ingersoll/round/alarm/moving head
	36	Bayard/wall clock
Popeye	34	New Haven/round/friends on outside

POCKETWATCHES

NAME	YR	NOTES
Betty Boop	34	embossed back
Big Bad Wolf	34	Ingersoll/embossed/winking eyes
Buck Rogers	35	embossed back/lightning bolt hands
Dizzy Dean	35	New Haven/with second hand
Donald Duck	39	Ingersoll/Mickey decal
Flash Gordon	39	New Haven
Lone Ranger	39	New Haven/with decal
Mickey Mouse	33	Ingersoll/short-stem/embossed back
	33	long-stem/Ingergoll/embossed back
	34	English/fat Mickey
	34	English/rat Mickey
	38	lapel/black/round/with decal
Moon Mullins	33	Ingersoll
Popeye	34	New Haven/with friends
	35	New Haven/no friends
Rudy Nebb	33	Ingraham
Skeezix	36	Ingraham
Smitty	36	New Haven
Three Little Pigs	39	prototype
Tom Mix	34	Ingersoll/embossed back

Chapter Three
The Postwar Years:
1946-1958

Between the years 1940 and 1946 no watches were produced because of the war. However, in the fall of 1946, Kelton watch introduced a small, rectangular Mickey Mouse watch. It was the head of Mickey on a post with the head rotating as the hour hand turned. This watch was made for only one year, and in 1947 Ingersoll, now U.S. Time, introduced a Mickey 1947 rectangular model. This watch is about three-quarters the size of the pre-war 1939 model, and has no second hand. All of the watches produced after World War II were made three-quarters the size of those made before the war due to the lack of material. U.S. Time made a round Mickey, also white-faced, with fluting on the case above the number 12 and below the number 6. As companion pieces, there were three Donald's made: a rectangular Donald with a blue face and silver case, a round Donald with a blue face, and a rectangular Donald with a gold-toned case. There was also a Mickey made in 1947 with a rectangular, gold-toned case. Neither the round Mickey nor the round Donald had an inner circle on its face. The significance of this will become apparent as we discuss the 1948 models.

Alice in Wonderland. These are the 1955, 1958, and 1950 models. The last two were made by U.S. Time which became Timex. The first was made by New Haven. The New Haven comes with an animated Mad Hatter. The 1958 model was packaged with a porcelain or plastic statue. The 1950 model was packaged in a plastic teacup. Size S.

Alice in Wonderland, 1950, U.S. Time. The box is decorated with pictures from the movie. This watch was packaged in a plastic teacup. *Courtesy of Jack Feldman.*

In 1948, to celebrate the 20th anniversary of Mickey Mouse, U.S. Time issued a series of ten watches known as the Birthday Series. They included Mickey, Donald, Daisy, Joe Carioca, Bongo, Pinnochio, Dopey, Jiminy Cricket, Pluto, and Bambi. Each one is a round watch, and on its dial is an inner and outer circle. On all the watches in this 1948 collection the characters are small enough to fit within the inner circle. In 1949 U.S. Time reissued the watches with luminous dials and luminous hands. The characters are larger in the 1949 version, and do not fit within the inner circle. We have many examples of advertisements showing that all ten pieces were repeated in 1949. However, only seven have been found! Interestingly enough, some collectors have claimed to have seen advertisements showing only seven. (Do Joe Carioca, Bongo, and Dopey really exist?) Also in 1949, they made non-luminous dials using the die-cuts that were used on the luminous faces. In other words, there is an entire series of non-luminous watches in which the figures are larger than the inner circle. (There are two exceptions, Pluto and Bambi, where the watch dials show only the images of their faces.) In 1949 they changed the case design so that there was fluting completely around the circumference of the case. Unfortunately, U.S. Time mixed and matched their cases which resulted in the factory placing 1949 dials in 1948 cases.

I will summarize all of the above confusion. There are ten 1948 Birthday Series, small figures that were meant to be placed in cases with fluting at the 12 and the 6. There are ten advertised luminous 1949 watches, but only seven have been found. In the 1949 set, the cartoon figure is larger than the inner circle except for Pluto and Bambi. There are seven non-luminous, large figure watches because Joe Carioca has not been found in the luminous model or the large model. Finally, the round, white Mickey, which is often called a 1948 model, is actually a 1947 model.

In the 1940s, various companies introduced comic character watches that were not produced before the war, such as Daisy Duck, Danny and the Black Lamb, Little Pig with the Fiddle, Snow White, Joe Palooka, Captain Marvel, Luey Duck, Porky Pig, Puss n' Boots, Blondie, Babe Ruth, Dale Evans, and Roy Rogers. The 1947 Captain Marvel came in a chrome-colored case. However, in 1989 boxes of these watches in pink-gold colored cases surfaced, which were believed to be new-old stock that had never been used. The setting for the story surrounding this mystery is in South America where there was great demand for the pink-gold colored cases. For whatever

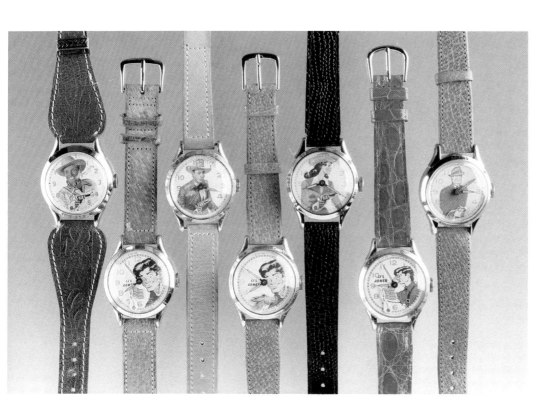

Animated watches, 1951. New Haven and Muros watches with animated moving guns or flags. There are three Li'l Abners, a Gene Autry, a Dick Tracey, a Texas Ranger and Annie Oakley.

Annie Oakley, made by Muros Watch Factory in 1950. It is a companion piece to the Paul Bunyan watch. Size M.

Ballerina, 1956, Ingraham. Size M.

reason, these watches were never shipped to South America and were recently discovered. Of course all of us collectors had to have one— especially since its makes such a great story.

In the late 1940s and early 1950s, New Haven introduced a watch with a moving apparatus on the lower part of the dial. There it was! —Dick Tracy and his automatic that moved at 120 beats per minute, Gene Autry and his revolver, and even a Dick Tracy with Gene Autry's revolver. Even though it was not the way the watch was meant to be assembled, it is believed that at the end of the day when the production line depleted their supply of Dick Tracy moving guns, the factory substituted the Gene Autry model. This watch is affectionately known as the Dick Tracy Cowboy. To complete the series of moving watches, there is the Texas Ranger, the recently discovered Paul Bunyan, Li'l Abner with a moving mule, Li'l Abner saluting a moving flag, and a Li'l Abner with a moving flag in which the character is looking at us. There is also a Cowgirl with a moving gun which collectors call Annie Oakley. I have recently discovered an advertisement with this designation and no longer consider the name of this watch to be questionable.

Bambi, made by U.S. Time as part of the Disney 1948 20th Birthday series consisting of ten watches. Notice the fluting above the 12 and below the 6. This is the advertised 1948 case, though it is sometimes sold in the 1949 case. Notice that there are two concentric circles; this is traditional on the 1948 and 1949 watches. Size M.

Bambi, made by U.S. Time in 1949. The outer circle is luminous so that it could be seen in the dark. The case is fluted all around the circumference of the watch. Only seven of the ten 1948 Birthday watches were made in the 1949 Luminous model. Size M.

Blondie, 1949, in box. *Courtesy of Jeff Cohen.*

Bongo the Bear, made by U.S. Time as part of the Disney 1948 20th birthday series consisting of ten watches. Notice the fluting above the 12 and below the 6. This is the advertised 1948 case. Bongo is smaller than the inner circle which is unlike the '49 series in which Bongo is larger than the inner circle. There are two concentric circles on both the 1948 and '49 series. Size M.

Blondie, 1949. Dagwood and animals are shown on hexagon shaped watch case. Copyrighted by King Features Syndicate. Size S.

Bugs Bunny, made in 1951 for sale at Rexall Drugstores. On this dial Bugs is holding a carrot in one hand and in the other hand he has a green bush from which two carrot hands extend and tell the time. Size M.

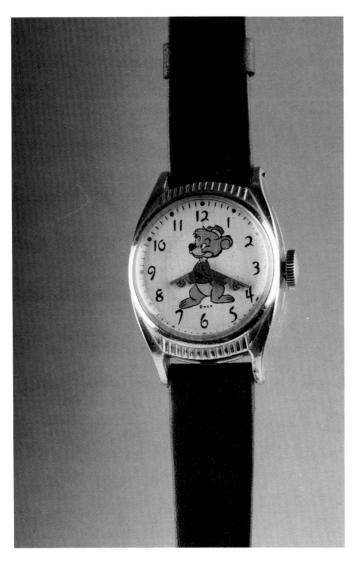

Bongo the Bear, U.S. Time, 1949. Bongo is larger than the inner circle on the dial in this 1949 version, unlike the '48 dial in which Bongo is smaller than the inner circle. The case, however, is from 1948, a variation that sometimes appears. Bongo was one seven watches from the 1948 Birthday series that reappeared in 1949. Size M.

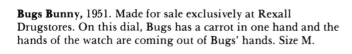

Bugs Bunny, 1951. Made for sale exclusively at Rexall Drugstores. On this dial, Bugs has a carrot in one hand and the hands of the watch are coming out of Bugs' hands. Size M.

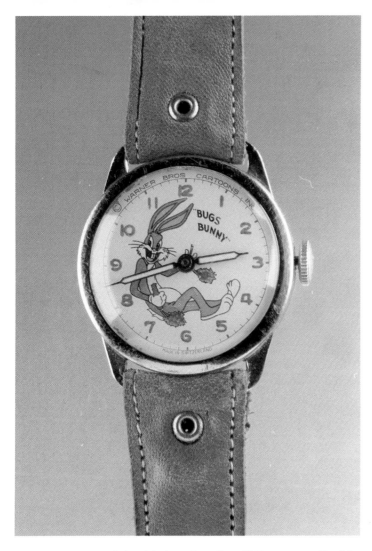

Bugs Bunny made in 1951 for sale at Rexall Drugstores. On this dial Bugs has a carrot in one hand and a stalk in the other from which come two green luminous hands. Size M.

Bugs Bunny, 1951, Ingraham. This clock features Bugs eating his carrot which moves up and down. *Courtesy of Jack Feldman.*

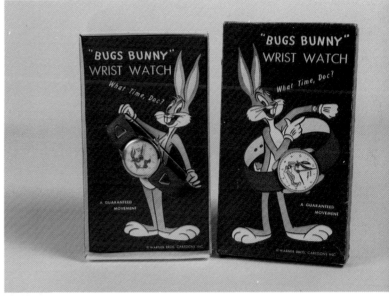

Bugs Bunny. A 1951 watch made for sale exclusively for Rexall Drugs. Notice that on the advertisement Bugs' hands are not carrot shaped as they are on the actual watch. Size M. *Courtesy of Jack Feldman.*

The 1950s brought us Cinderella, Alice in Wonderland, Shirley Temple, Robin Hood, and Woody Woodpecker. Other cowboys also appeared in the 50s: Roy Rogers, Gene Autry, Hopalong Cassidy, Lone Ranger, Davy Crockett, Red Rider, and the English cowboy Jeff Arnold. It was the beginning of the space age and we were introduced to many new space heroes such as Tom Corbett, Rocky Jones, Buzz Corey, and also the English Dan Dare. How could we forget Captain Marvel, Captain Marvel Jr., and Mary Marvel?

Take a close look at the Red Rider watch. If you use a loop and look at the center of the dial above Red's rifle butt, you will sometimes see a green bush. The Bugs Bunny watch, which was produced for sale only at Rexall Drugs, came in three models. The first and most famous has a bush in the middle, out of which come two carrots which serve as the hands. The second has a stalk in the center with green luminous hands. The third has nothing in the center, but the hands are orange to match the numbers around the dial. The same company that made Bugs also made the Red Rider. It appears that after making the initial run of Red Rider watches, the company had extra dials left over. In addition, many Bugs Bunny watches with the bush had not been sold. We assume that the factory placed the extra Red Rider dials on top of the already existing, unsold Bugs Bunny watches, resulting in many Red Rider watches with a green bush in the center.

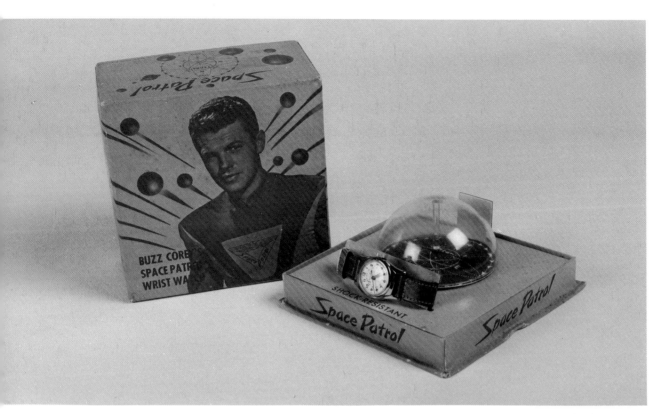

Buzz Corey, 1951, U.S. Time. Box is worth more than the watch. *Courtesy of Jack Feldman.*

Captain Liberty, a 1954 watch, its true value lies in the band. As with all watches made by Liberty, the name of the character is stamped on the watch. In this case the name stamped was Captain. Size M.

Captain Marvel, made by Fawcett, 1948. This watch came with three vinylite bands: red, green, and blue. It also came in a slightly smaller size with a one jewel movement, packed in a plastic box. Size M.

Captain Marvel, made by Fawcett, 1948. This is the famous (or is it infamous?) gold-tone Captain Marvel. The story is told that the case was made in pink-colored gold tone for sale in the South American market. Because of this, the watch was not advertised in the U.S. Size M.

Captain Marvel Jr., made by Fawcett, 1948. This is a companion piece to Mary Marvel and is considered extremely rare. Size S.

Cinderella. These are the Cinderellas that came out in 1950, 1955, and 1958. They were made by U.S. Time which became Timex. The 1950 model was packaged in a glass slipper. The 1955 model came both in a blue plastic case with a matching plastic band, and in a chrome colored case. The 1958 model came packaged with a porcelain or plastic statue. Size S.

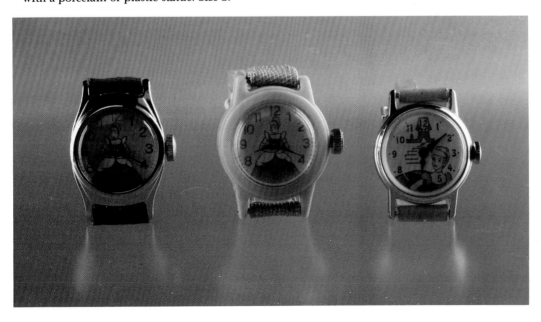

Cinderella, 1950. The plastic slipper box makes this packaging extremely rare. Size M. *Courtesy of Jack Feldman.*

Daisy Duck, made by Ingersoll, 1947. Was the forerunner of th round 1948 watch. This was the first female Disney character watch made. Size M.

One of the most heated discussion among collectors concerns the year that the first Minnie Mouse watch was made. Logic would claim that if Daisy were made in 1947, Cinderella and Alice in 1950, and an abundance of new female comic characters introduced in 1958, then Minnie would somehow have been produced in one of these time periods. However, the Disney Archives have no record of Minnie being produced prior to 1971! Even though the Minnie that many have called the 1958 model is similar in styling to the other 1958 watches, the guarantee sometimes found in the Minnie Mouse box has a return address which includes a zip code. Zip codes did not exist in 1958! Therefore, there was no Minnie Mouse watch made until 1971, even though a preponderance of collectors still believe that it is a 1958 Minnie. I will follow the tradition of my predecessors in another medium (who still claim Abner Doubleday invented baseball) and will show the watch in this chapter, but with an asterisk (*) noted.

In 1950 a round Mickey Mouse watch was introduced and repeated for eight consecutive years. Many of what are considered to be the most unappealing comic character watches made were issued in 1958, having only names on the dials with no caricatures. By the end of 1958, so many watches had been produced, approximately 25 million, that for all practical purposes production ceased. It was not to start again in quantity until ten years later.

During the period of 1946-1958, there only were seven pocket watches produced: Captain Midnight, Captain Marvel, Dick Tracy, Hopalong Cassidy, Donald Duck, Jeff Arnold, and Dan Dare. This was a time in which people no longer bothered to carry the pocket watch which had been so popular before World War II. To our knowledge, there was no Mickey pocket watch made during this period.

Clocks became more prevalent and more interesting than pocket watches. There was a Bugs Bunny electric, a Davy Crockett rocking horse, Hopalong Cassidy, various Mickeys, Pluto, Roy Rogers, and Woody Woodpecker.

If this period is noted for anything besides watches, it is for the packaging: Cinderella displayed in a glass slipper, Zorro on a felt hat, Mickey on a birthday cake, Davy Crockett on a powderhorn, Alice in Wonderland in a teacup, Hopalong Cassidy on a saddle, and Snow White and her magic mirror box, to name only a few.

To this collector, the watches of 1946-1958 are the most interesting. Of the pre-1972 eras, it has the widest variety of characters, the most unique selection of boxes/packaging, the most controversial stories, and the least amount of knowledge, creating in my mind the most fascinating period of watch production.

Daisy Duck, 1948, Ingersoll. Part of the 20th Anniversary edition of ten watches. Daisy is within the inner concentric circle, different from those made in 1949. There is fluting above the 12 and below the 6. This was the advertised case in 1948. Size M.

Daisy Duck, made by U.S. Time in 1949. Daisy is larger than the inner circle and the outer circle is luminous so that it could be seen in the dark. The case is fluted all around the circumference of the watch. Only seven of the ten 1948 Birthday watches were made in the 1949 luminous model. Size M.

Daisy Duck, made by U.S. Time as part of the Disney 1948 20th birthday series consisting of ten watches. This advertised 1948 case has fluting at the 12 and 6, but the Daisy on the dial is larger than the inner circle which dates it to 1949. On the 1948 production ones the characters were always smaller than the inner circle. All '48 and '49 models consisted of two concentric circles. Size M.

Dale Evans, made by Ingraham, 1951. This is a companion piece to Roy Rogers and also came in a round tonneau shape. Size M.

Dale Evans, made by Ingraham, 1949. This is the only one of Dale standing and was the first Dale Evans watch. Size M.

Dale Evans, made by Ingraham, 1955. Four models of the round tonneau shaped watch with matching bands and cases. Size M.

Danny and the Black Lamb, made by Ingersoll, 1947, and part of the series that included Snow White, Louie Duck, and Little Pig with Fiddle. Size M.

Davy Crockett made by U.S. Time, 1954. This is considered to be the first Davy Crockett Watch. Size S.

Danny and other Disney character watches advertised, 1947.

Davy Crockett, made by U.S. Time, 1954. The watch comes with a green plastic case, and is packaged on a plastic powder horn. Size S.

Davy Crockett, made by Ingraham, 1954. This watch came with a 3-D pop-up box. Size M.

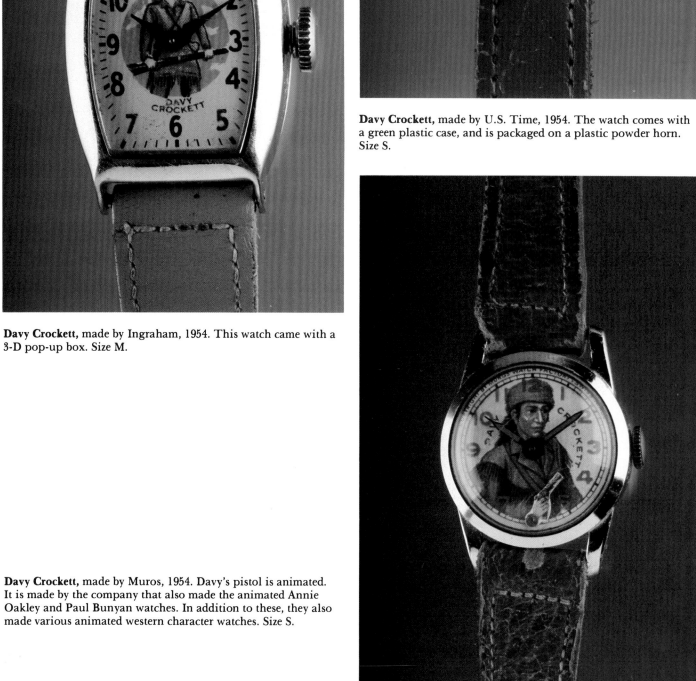

Davy Crockett, made by Muros, 1954. Davy's pistol is animated. It is made by the company that also made the animated Annie Oakley and Paul Bunyan watches. In addition to these, they also made various animated western character watches. Size S.

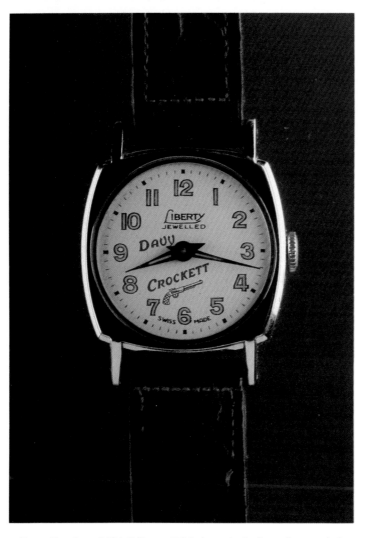

Davy Crockett, 1954, Liberty. This is typical of watches made by Liberty on which they stamped the character's name. Size M.

Dick Tracy, made by New Haven, 1948. This watch also came in a rectangular case. This uniquely styled case was also used on the Superman and Little Orphan Annie watches. Size M.

Davy Crockett, made by Ingraham from 1956-1962. they became Bradley in 1958. This must be a late production watch. This watch is noted for the uniqueness of its case. Size M.

Dick Tracy, 1951, New Haven. Western Dick Tracy. This watch has the western gun that is the same as the one on the Gene Autry watch. It is smaller in size than the one that has the automatic-style pistol. Many have appeared in this version, and it is assumed that these were made by New Haven. Size M.

Dick Tracy, 1948, New Haven. Size M.

Dick Tracy with animated automatic-style pistol, New Haven, 1951. Part of the series that included three Li'l Abners, Annie Oakley, Texas Ranger, and Gene Autry. Size M.

Disney 20th Birthday Series, 1948, made by Ingersoll. Birthday Series consisting of Pinnochio, Dopey, Jiminy Cricket, Bongo the Bear, Mickey Mouse, Donald Duck, Pluto, Daisy Duck, Bambi, Joe Carioca.

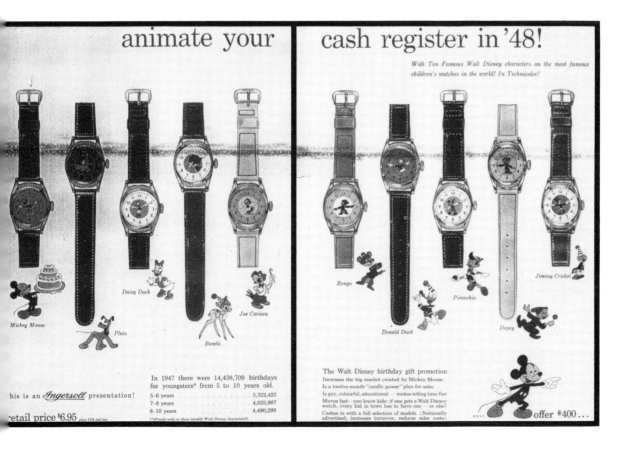

animate your cash register in '48!

With Ten Famous Walt Disney characters on the most famous children's watches in the world! In Technicolor!

Mickey Mouse   Daisy Duck   Pluto   Bambi   Joe Carioca

Bongo   Donald Duck   Pinocchio   Dopey   Jiminy Cricket

In 1947 there were 14,438,709 birthdays for youngsters* from 5 to 10 years old.

5-6 years	5,322,423
7-8 years	4,625,987
8-10 years	4,490,299

*(*already sold on these lovable Walt Disney characters*)*

this is an *Ingersoll* presentation!

etail price $6.95 *plus 10% fed tax*

The Walt Disney birthday gift promotion
Increases the big market created by Mickey Mouse.
Is a twelve-month "candle power" plan for sales.
Is gay, colourful, educational — makes telling time fun
Moves fast — you know kids: if one gets a Walt Disney
watch, every kid in town has to have one — or else!
Cashes in with a full selection of models. (Nationally
advertised, increases turnover, reduces sales costs)

offer #400 ...

Disney Character watches advertisement, 1948.

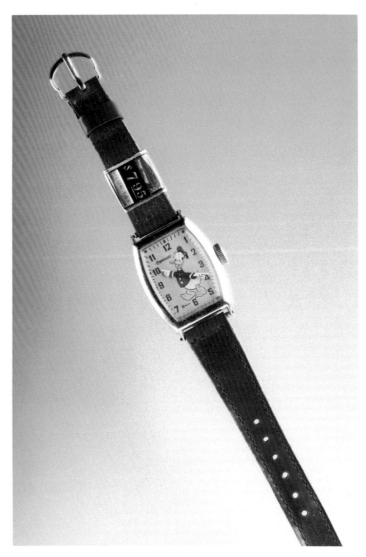

Donald Duck, 1947, Ingersoll. This is the gold case, deluxe model. Size M.

Donald Duck, made by Ingersoll, 1947. This watch is a companion piece to the round blue Donald Duck, and also the round and rectangular Mickey Mouse. This watch was also made in a deluxe gold case. Most models had Ingersoll printed on the dial, however there were models of Mickey made without the name Ingersoll. Size M.

Donald Duck, 1949. First made by U.S. Time as part of Disney's 1948 20th Birthday series consisting of ten watches. The case is the 1948 version with fluting above the 12 and below the 6, Donald is larger than the inner circle, meaning that this was the dial that was produced in 1949. There are two concentric circles on both the '48 and '49 series watches. Size M.

Donald Duck, made by U.S. Time in 1955. This watch was packaged in a paper pop-up box. Size S.

Donald Duck, made by U.S. Time in 1949. Donald Duck is larger than the inner circle and the outer circle is luminous so that it could be seen in the dark. The case is fluted around the circumference of the watch. Only seven of the ten 1948 Birthday watches were made in the 1949 luminous model. Size M.

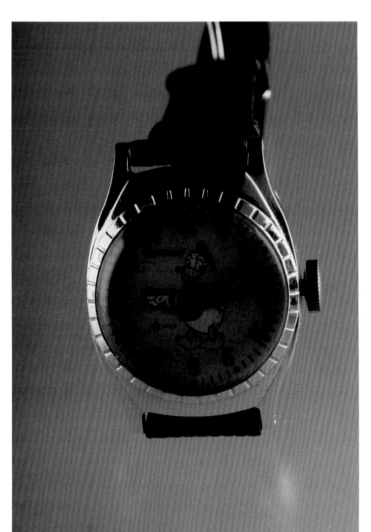

Donald Duck, 1947, Ingersoll. This is the companion piece to the 1947 Mickey Mouse. Notice that there is no circle on the '47 series. This case was not used in the 1948 Birthday series, but was used again in 1947. Size M.

Dopey, made by U.S. Time as part of Disney's 1948 20th Birthday series consisting of ten watches. This is the advertised 1948 case with a 1949 dial where Dopey is larger than the inner circle. There are two concentric circles on both the '48 and '49 series watches. Size M.

Dopey, made by U.S. Time as part of Disney's 1948 20th birthday series consisting of ten watches. This is the advertised 1948 case with fluting at the 12 and 6. Notice that the full figure of Dopey fits within the inner concentric circle unlike the 1949 model on which the character was larger than the inner circle. All '48 and '49 models consisted of two concentric circles of some fashion. Size M.

Gene Autry, made by Wilane, 1948. This is considered to be the first Gene Autry watch. Size L.

Gene Autry, 1948, by New Haven. Rumor has it that the first production run showed Gene Autry weighing a lot more. This picture shows him much thinner. Size M.

Gene Autry, made by New Haven in 1948. This is the same face that was used on the moving gun Gene Autry of 1951. Size M.

Gene Autry, made by New Haven, 1951. It has an animated gun which shoots 120 times per minute. Part of a series that included Li'l Abner, Dick Tracy, Texas Ranger, and Annie Oakley. Size M.

Hoky Poky, 1949, made by Acco. The hand is animated. Size M.

Hopalong Cassidy, made by U.S. Time in 1955. This is a plastic version of the 1950 watch and also came packaged in the infamous saddle. Size S.

Hopalong Cassidy, made by Ingersoll in 1950. It came in two sizes, small and medium. The original medium size sold for two dollars less than the smaller size. The packaging was a paper saddle and the watch was wrapped around it.

Hopalong Cassidy, made in 1960 in Great Britain. The only way to tell this watch apart from its American counterpart (besides taking it apart), is to look at Hoppy's collar which is more horizontal than the American version. Size S.

Hopalong Cassidy 1950, U.S. Time. The saddle stand box sold for $7.95. Size M. *Courtesy of Jack Feldman.*

Hopalong Cassidy, 1950, U.S. Time. Size M. *Courtesy of Jack Feldman.*

Howdy Doody, made by Patent Watch Company 1954. Howdy's eyes move with the hour hand. The watch came in two sizes: medium and small.

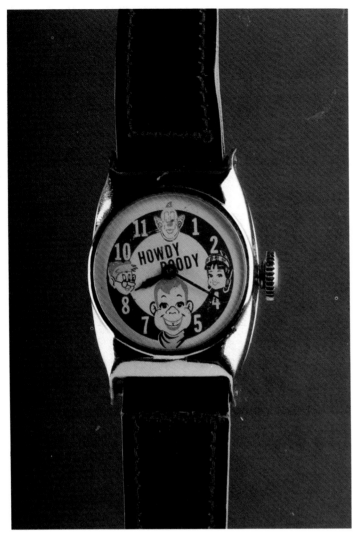

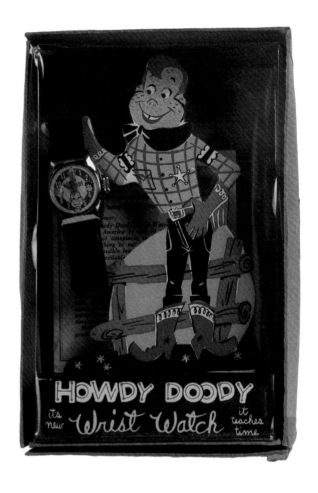

Howdy Doody made by Ingraham, 1954. Howdy and his friends came in a stand-up paper box. Size S.

Howdy Doody, 1954, Ingraham. The unique display box is extremely rare. *Courtesy of Jack Feldman.*

Ingersoll, 1947. These are the four watches that are most difficult to find of this period. They are Snow White, Little Pig with Fiddle, Danny and the Black Lamb, and Louie Duck.

Jeff Arnold, 1953, Ingersoll. English cowboy. Jeff's arm is animated and moves up and down. *Courtesy of Roy Ehrhardt.*

Mickey Mouse birthday watch advertisement, 1948. Notice not only the watch and the pen, but also the sterling silver ring.

Jiminy Cricket, made by U.S. Time as part of the Disney 1948 20th birthday series that consisted of ten watches. The fluting on the case and the size of Jiminy on the dial are both characteristic of the 1948 model. Size M.

Jiminy Cricket, made by U.S. Time in 1949. Jiminy is larger than the inner circle and the outer circle is luminous so that it could be seen in the dark. The case is fluted around the circumference of the watch. Only seven out of ten of the 1948 Birthday series watches were made in this fashion in 1949. Size M.

Jiminy Cricket, 1948, Ingersoll. Part of the 20th Anniversary Birthday Series, there were ten watches to the set. Jiminy is larger than the inner circle, meaning this is a '49 dial in the '48 case, actually made this way by the factory. The '48 figure was always smaller than the inner concentric circle. Size M.

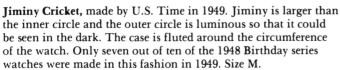

Jiminy Cricket. One of the ten Birthday Series watches to celebrate the 20th anniversary of Mickey's birthday. All ten watches came in identical cases. Size M. *Courtesy of Jack Feldman.*

Joe Carioca, made by U.S. Time as part of Disney's 1948 20th Birthday series consisting of ten watches. Size M.

Joe Palooka, 1947. Made by New Haven. This is the only Joe Palooka watch ever made. Size M.

Joe Carioca, 1953. This is a variation of the character that was on the 1948 and 1949 watches. Size M.

Junior League, 1956, Ingraham. A companion piece to Junior Nurse. Size M.

Li'l Abner, with an animated flag, 1951, New Haven. Part of a series of three Li'l Abners, Dick Tracy, Gene Autry, Annie Oakley, and the Texas Ranger. Size M.

Junior Nurse, 1956, Ingraham. A companion piece to Junior League. Size M.

Li'l Abner, with an animated mule, 1951, New Haven. Part of a series of three Li'l Abners, Dick Tracy, Gene Autry, Texas Ranger, and Annie Oakley. Size M.

Li'l Abner, with an animated flag, 1951, New Haven. Part of a series of three Li'l Abners, Dick Tracy, Gene Autry, the Texas Ranger, and Annie Oakley. Size M.

Little Pig with Fiddle, made in 1947 by Ingersoll. Part of the series that included Snow White, Louie Duck, and Danny and the Black Lamb. Size M.

Li'l Abner, made in the mid-50s and very rare. Size S.

Lone Ranger made by New Haven, 1948. This was a reissue of the 1939 Lone Ranger but in a smaller size. Size M.

Lone Ranger, made in 1951. Rare and difficult piece to find. Size S.

Louie Duck, made by Ingersoll in 1947. This was advertised in a group of four watches, all of the same shape. All are extremely rare. The others in the series are Danny and the Black Lamb, Little Pig with Fiddle, and Snow White. As you can tell, only one of these survived as a character on its own. Size M.

Lucy, 1958, Bradley. This watch is often confused with the 1974 Bradley. Size S.

Mary Marvel Jr., made by Fawcett, 1948. Made as a companion piece to Captain Marvel Jr. Size S.

Mickey Mouse, made by Kelton, 1946. One of the first watches made after WWII. Mickey's head is on a post that rotates with the hour hand. The watch was only made for one year and came with a case variation and a different numbering system on the dial. Size M.

Mickey Mouse made by Ingersoll in 1947. It came in a silver case and in a deluxe gold-tone case. While the silver case is readily available, the opposite is true of the gold. The word Ingersoll appears under the number 12, as it also does on the round version of this piece. At the same time there were matching Donald Duck round and rectangular watches made with a blue background. Size M.

Mickey Mouse, made by Kelton, 1946. Another of the first watches made after WWII. Mickey's head is on a post that rotates with the hour hand. This watch, too, was only made for one year and came with a case and dial variation. Size M.

Mickey Mouse, made by Ingersoll. From left to right: the '47 Mickey, the '48 Mickey in a '48 case, the '49 Mickey in a '48 case, and the '49 Mickey luminous in a '49 case. Size M.

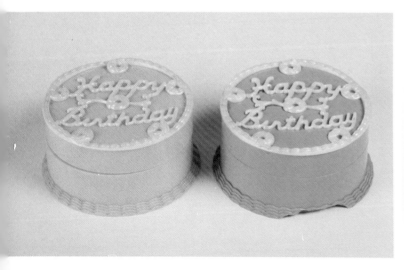

Mickey Mouse birthday cakes for 1947 Mickey and Donald Ingersoll watches. Size M. *Courtesy of Jack Feldman.*

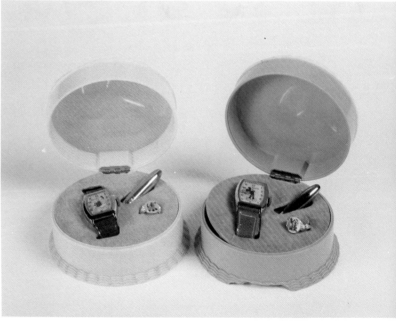

Mickey Mouse open birthday cakes for 1947 Ingersoll watches Mickey and Donald showing pen and ring that came with the watch. Size M. *Courtesy of Jack Feldman.*

Mickey Mouse made by Ingersoll, 1947. This watch was really the first mass produced watch issued after WWII. The dial sometimes had the name Ingersoll on it. It was also made in a deluxe model with a gold-tone case. Its companion pieces are a rectangular and a round Donald Duck. Size M.

Mickey Mouse, Ingersoll, 1948 20th Anniversary Birthday Series. One of ten watches in the series. Mickey is smaller than the inner circle unlike the '49 where he is bigger than the inner circle. There are two concentric circles on both the '48 and the '49. There is fluting above the 12 and below the 6 as was advertised in the 1948 models. Size M.

Mickey Mouse, 1949, U.S. Time. Plastic case table clock. *Courtesy of Jack Feldman.*

Mickey Mouse, 1952, U.S. Time. A companion piece to Minnie Mouse. Size M. *Courtesy of Jack Feldman.*

Mickey Mouse alarm and wristwatch advertisements, 1947. Notice that there are no concentric circles on the Mickey and Donald watches.

Mickey Mouse, the 1950, 1958, and 1955 models. The 1950 was made by Ingersoll, which became U.S. Time, which became Timex. The 1950 model was the last watch made that had the name Ingersoll on the dial. It came packaged in a flat box, similar to the 1947 model. It was reissued in 1952, packaged in a stand-up paper Mickey. The 1958 model is the same and was packaged in a porcelain or plastic statue. The 1955 red plastic came originally with a plastic band. Size S.

Mickey Mouse, 1955, U.S. Time. Mickey Mouse and Donald Duck pop-up display boxes are unique. Size M. *Courtesy of Jack Feldman.*

Mickey Mouse. A 1958 store display for Mickey and Minnie watches. *Courtesy of Jack Feldman.*

Mickey Mouse. Another 1958 store display for Mickey and Minnie watches. *Courtesy of Jack Feldman.*

Minnie Mouse, 1958, U.S. Time. A companion piece to Mickey Mouse. Size M. *Courtesy of Jack Feldman.*

Orphan Annie, 1948, New Haven. Part of a series that included Superman and Dick Tracy. Size M.

Orpan Annie, made by New Haven, 1948. It also came in a rectangular case. This unusual case was also used on Superman and Dick Tracy watches. Size M.

Orphan Annie, 1948, New Haven. This is a watch that is smaller than the 1935 model, but the packaging is similar to all the '48 packages. Size M. *Courtesy of Jack Feldman.*

Pinnochio. Made by Ingersoll as part of the Disney 1948 20th Birthday Series consisting of ten watches. Notice the fluting above the 12 and below the 6. This is the advertised 1948 case. Size M.

Pinnochio, made by U.S. Time in 1949. Notice that in 1949 Pinnochio is larger than the inner circle and that the outer circle is luminous so that it could be seen in the dark. The 1949 case is fluted all around the circumference of the watch. Size M.

Pinnochio, made by U.S. Time, originally as part of Disney's 1948 20th birthday series consisting of ten watches. This is in the advertised 1948 case, but Pinnochio is larger than the inner circle and this was the dial that was produced in 1949. There are two concentric circles on both the '48 and '49 watches. Size M.

Pluto. This is one of the 1949 Birthday series Disney watches made by U.S. Time. It consists of two concentric circles as does the '48 birthday series, however the outside circle is luminous—so it shines in the dark. The case is also different because it has fluting completely around the circumference of the dial. Even though there are advertisements for all ten '48 birthday series watches being shown as luminous models, only seven characters were actually made. Size M.

Pluto, made by U.S. Time as part of the 1948 Disney's 20th birthday series that consisted of ten watches. Notice the fluting above the number 12 and below the number 6 on the case. This is the true advertised 1948 case and dial. Size M.

Popeye, manufacturer unknown, 1948. Neither the watch nor box is copyrighted, but this watch was a production item. Size M.

Popeye, made in 1948 by an unknown watch company, it has the same case design as the Dagwood and Blondie watch of the same period. This is not the Popeye as we know it and no reason has been found for this version. Size S.

Porky Pig, made by Ingraham, 1949. There was a round version of this watch made in 1951. Size M.

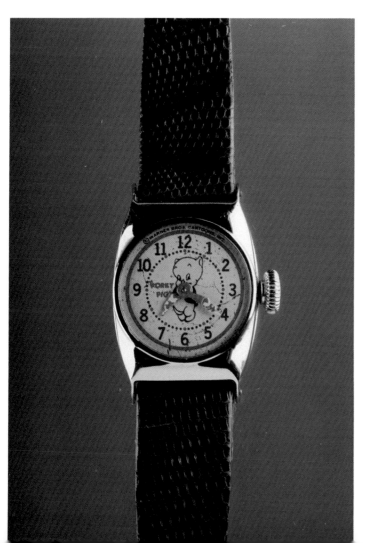

Porky Pig, made by Ingraham, 1949. This is the first Porky Pig watch. Size S.

Punkin Head, made by Ingraham, 1947. This watch is a prototype and never went into production. Size M.

Red Rider, 1951, made by the same company that made the Bugs Bunny watch that was only sold in the Rexall Drugstores. This paper dial was pasted over the Bugs Bunny dial. If one looks closely at the very center of the watch where the hands meet, behind Red Rider's rifle butt you can see the green bush that was the center of the Bugs Bunny watch. Size M.

Robin Hood, made by Bradley, 1956. This watch came in a pop-up 3-D box. Size M.

Puss-N-Boots, 1949, made by Nuhope. Watch should have animated cat hands. Size S.

Robin Hood, made by Viking, 1958. This watch was always considered to be made prior to WWII because of its large size. But recent advertisements have been found that properly date it to this period.

Rocky Jones, made in 1954 by Ingraham. The Rocky Jones Space Ranger watch came with a space ranger band which is also highly collectible. Size M.

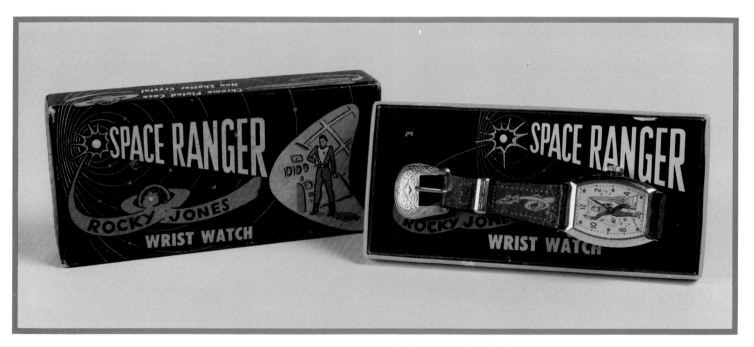

Rocky Jones, 1954, by Ingraham. The design on the band makes this watch even more valuable. Size M. *Courtesy of Jack Feldman.*

Roy Rogers, made by Ingraham, 1951. This is considered to be the first Roy Rogers watch. Size M.

Roy Rogers, made by Ingraham, 1951. Round tonneau. This is a companion piece to the rectangular watch. Size M.

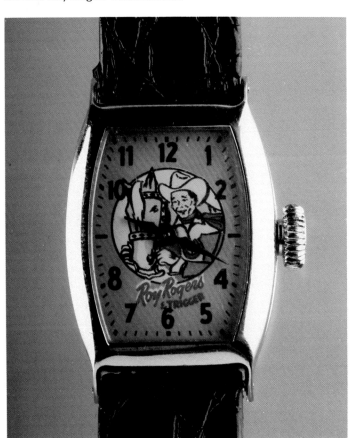

Roy Rogers, made by Ingraham in 1951. The watch came with an expansion metal bracelet or with a black leather strap. It also came in a round tonneau case. Size M.

Roy Rogers, made by Bradley Time, 1955. Companion piece to Dale Evans. Size M.

Roy Rogers, made by Ingraham, 1951. This is a Roy Rogers and Trigger reflector watch and is a companion piece to Dale Evans. Size M.

Snow White, made by Ingersoll, 1947. One of the series of four which included Danny and the Black Lamb, Little Pig with Fiddle, and Louie Duck. In 1948 this watch was produced in a round version. The only one of the four characters to ever be repeated. Size M.

Rudolf the Red Nose Reindeer, 1947, Ingersoll. Size M.

Snow White. These are the 1958, 1950, and 1955 models. The 1958 Snow White and Dopey were packaged in a porcelain or plastic statue. The 1950 watch was packaged in a magic mirror box. The 1955 watch with the matching yellow plastic band was packaged with a molded plastic figure. Size S.

Space Explorer, 1954 compass watch. This watch sold for $5.95. Size M.

Space Explorer, 1954 model. Size M.

Space Patrol, made by U.S. Time. This was packaged with a plastic dome and has 13-24 hour markers. Size S.

Superman, made by New Haven, 1948. This watch also came in a rectangular case. This unusual case design was also used on a Dick Tracy watch and a Little Orphan Annie watch. The Superman band is also highly collectible. Size M.

Superman, made by New Haven in 1955. This watch is noted for its green background and also its bolt action hands. It is also the first Superman watch to show his full body. Size M.

Texas Ranger, Left: 1951, New Haven. This was part of the series that included Annie Oakley, Dick Tracy, Gene Autry, and three Li'l Abners. Size M. Right: 1971, Sheraton. A reissue in a small size with slight variations in the artwork.

Tom Corbett, made by Ingraham, 1951. The display card that this watch came with is a paper spaceship. Size S.

Woody Woodpecker, 1950, Ingraham. This is considered to be the first Woody. Size S.

Tom Corbett, 1951, Ingraham. The display card is as important as the watch. *Courtesy of Jack Feldman.*

Woody Woodpecker, made by Ingraham, 1950. This is the first Woody Woodpecker watch. Size M.

Woody Woodpecker 1950 store display for Woody Clocks. *Courtesy of Jack Feldman.*

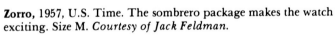

Zorro, 1957, U.S. Time. The sombrero package makes the watch exciting. Size M. *Courtesy of Jack Feldman.*

Zorro, 1950, Ingersoll. The value lies in the strap and the packaging, which was a copy of Zorro's hat. Size M.

THE POSTWAR YEARS: 1946-1958

WRISTWATCHES

NAME	YR	NOTES
Alice in Wonderland	50	blue backgrnd/silver case
	54	pink case and band
	58	coming out of flower
	58	Timex/name only
Annie Oakley	51	moving gun
Ballerina	56	Bradley/animated feet
Bambi	48	birthday
	49	luminous
Blondie	49	Dagwood & animals
Bongo	48	birthday/large figure
	48	birthday/small figure
Bugs Bunny	51	bush in ctr/carrot hands
	51	carrot ctr/green #/hands
	51	no ctr/org #/blade hands
Captain Liberty	54	embossed airplanes on band
Captain Marvel	48	pink gold
	48	round silver
Captain Marvel, Jr.	48	tonneau
Cinderella	50	pink backgrnd/blue plastic case
	50	pink backgrnd/metal case
	58	castle at 12
	58	Timex/name only
Daisy Duck	47	rectangular
	48	birthday/large figure
	48	birthday/small figure
	49	luminous
Dale Evans	49	large size
	49	rectangular/standing
	57	blue tonneau
Dale Evans	57	gold tonneau
	57	pink tonneau
	57	rectangular
	57	silver tonneau
Danny & the Black Lamb	47	red rectangular
Davy Crockett	51	moving knife/Muros
	54	round silver
	54	small, round/green plastic
	54	with gun/lettering by Liberty
	56	standing yellow/rectangular
	56	tonneau/yellow
Dick Tracy	48	rectangular
	48	tonneau
	51	moving gun
Donald Duck	47	blue face/round/no circle
	47	blue face/rectangular silver case
	47	rectangular gold
	48	birthday/large figure
	48	birthday/small figure
	49	luminous
	50	round/silver/small
	58	Timex/name only
Dopey	48	birthday/large figure
	48	birthday/small figure
Gene Autry	48	rectangular/on Champion
	48	round face
	51	moving gun
Hoky Poky	49	moving card hand
Hopalong Cassidy	50	large
	50	small, black plastic
Howdy Doody	54	large/moving eyes
	54	tonneau w/friends
Jiminy Cricket	48	birthday/large figure
	48	birthday/small figure
	49	luminous
Joe Carioca	48	birthday/large figure
	48	birthday/small figure
	53	animated hands
Joe Palooka	47	small/rectangular
Junior League	56	Bradley
Junior Nurse	56	Bradley
L'il Abner	50s	small/silver/black & white face
	51	moving flag/front view
	51	moving flag/side view/salute
	51	moving mule
Little Pig	47	with fiddle
Lone Ranger	47	rectangular
	51	round Silver
Louie Duck	47	rectangular
Lucy	58	Bradley/writing under 6
Mary Marvel	48	small size
Mickey Mouse	46	Mickey Kelton/head only/gold face
	46	Mickey Kelton/head only/white face
	47	gold rectangular
	47	round/white-face
	47	silver rectangular
	48	birthday series/small figure
	48	birthday/large figure
	49	luminous
	50	round/silver/small
	50	round/silver/small/Ingersoll under 1
	58	red plastic
	58	Timex/name only
Minnie Mouse	58	round/silver
Orphan Annie	47	small/rectangular
	48	tonneau
Pinnochio	48	birthday/large figure
	48	birthday/small figure
	49	luminous
Pluto	48	birthday
	49	luminous
Popeye	48	unknown/friends on face
	49	hexagonal case
Porky Pig	49	rectangular
	49	round
Punkin Head	48	Ingraham/rectangular
Puss 'n Boots	47	Saro
Red Rider	49	round
Robin Hood	56	rectangular
	58	round
Rocky Jones	54	rectangular
Roy Rogers	51	on Trigger
	54	sitting on Trigger/facing right
	55	large face/yellow
	57	rectangular/green
	57	tonneau/green
Rudolph the Red Nose Reindeer	47	Ingersoll
See-saw Marjorie Daw	54	small size
Snow White	47	rectangular
	48	Ingersoll/round
	50	white backgrnd/silver
	58	Timex/name only
	58	Timex/with Dopey
	58	white backgrnd/yellow plastic
Space Explorer	54	compass in middle
	54	yellow rocket ship
Space Patrol	50	small/silver
Superman	47	rectangular
	48	tonneau
	55	bolt-hands/green face
Texas Ranger	51	Muros
Tom Corbett	51	tonneau
Toppie	51	Ingraham/tonneau
Westerner	54	Ingraham
Woody Woodpecker	50	rectangular
	50	tonneau shaped
Zorro	50	black face

CLOCKS

NAME	YR	NOTES
Bugs Bunny	51	electric clock
	51	Ingraham/alarm
Davy Crockett	54	Hadden/horse bucks
	55	Pendulette
Donald Duck	50	Glen/moving head
Hopalong Cassidy	50	U.S.Time/black/round
Lady	55	Allied/plastic/figure
Mickey Mouse	47	Ingersoll/plastic/alarm/luminous hands
	47	Ingersoll/round/metal case
	49	Ingersoll/plastic/electric
	49	round/thick metal
	55	Allied/plastic figure
Pluto	53	Allied/plastic/figure
Roy Rogers	51	Ingraham/moving horse
Schmoo	47	New Haven/white plastic/figure
Woody Woodpecker	50	Columbia/wall clock
	50	Columbia/Woody's Cafe

POCKET WATCHES

NAME	YR	NOTES
Captain Marvel	48	New Haven
Captain Midnight	48	Ingraham
Dan Dare	53	Ingersoll/double animation
Dick Tracy	48	Ingersoll
Donald Duck	54	Swiss-made
Hopalong Cassidy	50	U.S. Time/black/round
Jeff Arnold	53	Ingersoll/animated pistol
Peter Pan	48	Ingraham

Chapter Four
From Bust to Boom:
1958-1972

By the end of 1958, so many watches had been produced that for the next ten years production almost ceased. Bradley Watch, a division of Elgin Industries, was the only American watch company who continued to produce watches. They made a Pinnochio, Quick-Draw McGraw, Yogi Bear, and Popeye, all of which were small and round and meant to be worn only by children. There was no demand for Disney watches and, therefore, none were produced in this country during this time. However, the Japanese began producing their version of American comic characters in addition to their own characters such as Astroboy and Godzilla. In 1968 when Sammy Davis, Jr. and Johnny Carson began wearing the Timex Mickey Mouse watch, the craze began again.

Timex had decided that it might be the right time to reissue a comic character watch. They introduced a series of three: Mickey Mouse, Minnie Mouse (they sold for $12.95 each), and the first battery operated Mickey watch which sold for $19.95. From 1968 until the end of 1972, various companies were licensed to produce Mickey Mouse watches, many sold exclusively at Disneyland. There were three other electric Mickeys made during this period: one by Hamilton, one by Elgin, and one by Helbros. Vantage Watch Company made watches that were to be sold only at the park, two of which have become extremely collectible. One model has a clear case in the back where the movement can be seen through the case. The second collectible model shows Mickey with white hands, an apparent mistake that the factory corrected when it was discovered. However, approximately 500 are thought to have survived.

Between 1968 and 1972 other watch companies entered the marketplace. Sheffield Watch produced a series of large, over-sized watches including Popeye, Felix, Daffy Duck, and Merlin the Magic Mouse. Because of the detail and large size, these watches have become highly desirable. Sheraton Corporation commissioned a series of watches which consisted of Cool Cat, Porky Pig, Daffy Duck, Elmer Fudd, and Wile E. Coyote. It was during these years that the revolving disc watches were introduced by Rouan Watch Company: Archie, Peter Pan, and Wendy. Prince Roable Watch Company gave us Fred Flintstone, Pebbles, and BamBam. In the mid-1960s, Gilbert Watch Company produced a series of plastic watches such as the Bronco, James Bond, and G.I. Joe. This period brought us the onslaught of many advertising wristwatches, such as Planters Peanut and Charlie the Tuna, only to name a few.

In late 1971 Helbros Watch Company introduced a Mickey watch with a 17-jewel movement along with companion pieces like Minnie Mouse, and, what I consider to be the most comical of all comic character watches, the first backwards Goofy! The Goofy watch was priced at $19.95, which was a horrendously high price for a comic character watch at that time. Its most unusual features are the counter-clockwise numbers and the hands running

Allstar Baseball, 1966, autographed by Mickey Mantle, Roger Maris, Willie Mays. This watch came with a green dial which is more scarce than the black dial. This is a companion piece to Allstar Football. Size M.

Allstar Basesball, 1966, black face. Size M.

backwards which, unfortunately, made it difficult to teach children to tell time. In its day, it was not very successful; however, today it is considered one of the most valuable comic character watches and has sold for $1,800.

My favorite watches from this period were the one introduced by the Jay Ward Studio. There were 16 to the series including Rocky, Bullwinkle, Boris, Natasha, Dudley DooRight, George of the Jungle, Snidely Whiplash and his own caricature watch. Jay Ward himself had favorites such as Bullwinkle and Dudley DooRight in which the dials featured painted scenes of the characters.

In late 1971 Timesetters introduced the first Tweety and Sylvester watches, and Hi Time gave us Smokey Stover. In the early 1970s, Robert Lesser, author of the *Celebration of Comic Art and Memorabilia,* became involved in both Huckleberry Time and the Precision Time Company. Huckleberry had an interesting concept: take a pocket watch, put lugs on the back, put a band through the lugs and wear the pocket watch as a wristwatch. The most famous of these is the Buck Rogers, but there were others such as Mickey Mouse, Valentino, Marilyn Monroe, and Charlie Chaplin.

The history of Precision Time—which consisted of only one watch—is even more interesting. The company involved a joint venture with Japanese investors to produce comic character watches. The first watch chosen for production was Flash Gordon. At a meeting to review the first production run, it became apparent that there had been a large breakdown in communications. The investors thought the company was producing Mickey Mouse watches and had absolutely no idea who Flash Gordon was. Production stopped with the Flash Gordon model as the only watch made by Precision Time.

In 1972 Elgin Watch Company assigned the rights to make comic character watches to their Bradley subsidiary—and a new era of comic character watches began. No longer would we have just a few models a year to choose from. Instead they introduced 20 to 30 different designs for each character. In my mind, this concluded the "golden years" of comic character watches.

Roy Rogers and Superman were the only pocket watches known to have been made during this period. However, there was an interesting series of clocks produced in the mid-1960s by the Bayard Company of France, on which the characters' heads wagged. They were Donald Duck, Mickey Mouse, Pinnochio, Pluto and Snow White. The Beatles Yellow Submarine clock and the Popeye with the animated Sweet Pea are also highly desirable.

This period of comic watch collecting is interesting because of the entry of new characters by Jay Ward, the over-sized series of the Sheratons, the first 17-jewel and electric models, and the most unique backwards Goofy. This period did not give us a great quantity of watches, but each one is interesting in its own right.

Allstar Football, 1966, autographed by Mike Ditka, Jim Brown, and Jim Taylor. This is a companion piece to the Allstar Baseball watch. Size M.

Andy Panda, 1972, by Endora Time. This is the first Andy watch. Size M.

Andy Panda. This watch was made in Japan during the mid-60s and had a clear back. Size M.

Andy Panda, 1971, Japanese model. Size M.

Archie, 1972, Rouan. Features a revolving disc. Part of the series that included Wendy, Peter Pan, and Woody Woodpecker. Size M.

Ballerina, Bradley, 1962. Size M.

Astro Boy, 1966. This Japanese superhero watch was made in Japan for sale only in that country. It is part of a series of six different Japanese characters.

Ballerina, 1964, Bradley. Size M.

BamBam, made by Prince Roable, 1971. Part of the series that included Pebbles and Fred Flintstone. Size S.

Bambi, 1964, Bayard. A series of clocks produced in France with limited distribution in the U.S. Others in the series include Pluto, Mickey Mouse, Donald Duck, Snow White, and Pinnochio. The heads of the characters wag back and forth. *Courtesy of Jack Feldman.*

Barbie, 1963. This is the only watch series that has the year the watch is made on the dial. Size S.

Barbie, 1964. Size S.

Barbie, 1971. Size S.

Barbie, 1964. Size S.

Barbie and Ken, 1964. Size M.

Baseball Player, 1960, Ingraham reflector. Size M.

Boris, made for Jay Ward in 1971. This is a seventeen-jewel movement watch and is part of a series of sixteen watches. Size M.

Batman, 1966, Gilbert. Other plastic molded watches like these include James Bond, GI Joe, and the Western Bronco. Size M.

Bozo the Clown, 1970. Issued in conjunction with Capitol Records. Size M.

Buck Rogers, 1971, Huckleberry Time. This pocket watch could also be worn as a wristwatch. This watch is made from pick-up art taken from the packaging of the 1935 pocketwatch. This is easily discernable because Wilma is facing left whereas on the original pocketwatch she was facing front. Size L.

Brutus, 1966. A Japanese-made watch, not for sale in the U.S. Size M.

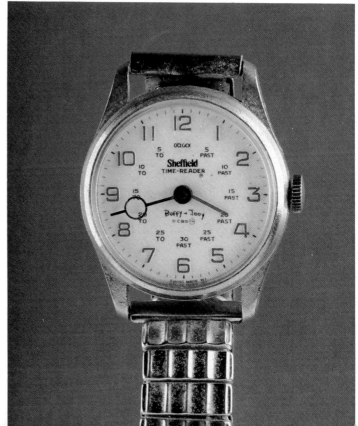

Buffy and Jody, 1971, Sheffield. This "Time-Reader" was designed to help children learn to tell time. Size M.

Bugs Bunny, a 1960s Japanese version of Bugs. Size M.

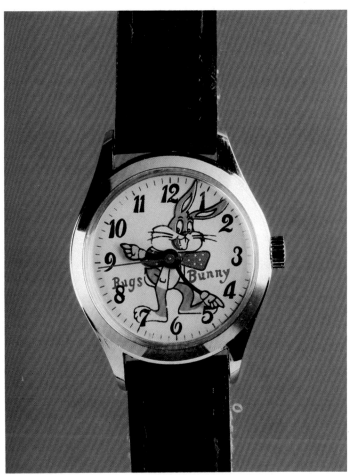

Bugs Bunny, 1972. A Hong Kong version of Bugs. Size M.

Bugs Bunny, 1971, Japanese Bugs Bunny. Size M.

Bugs Bunny. The 1972 version of Bugs's face. Made in Japan. Size S.

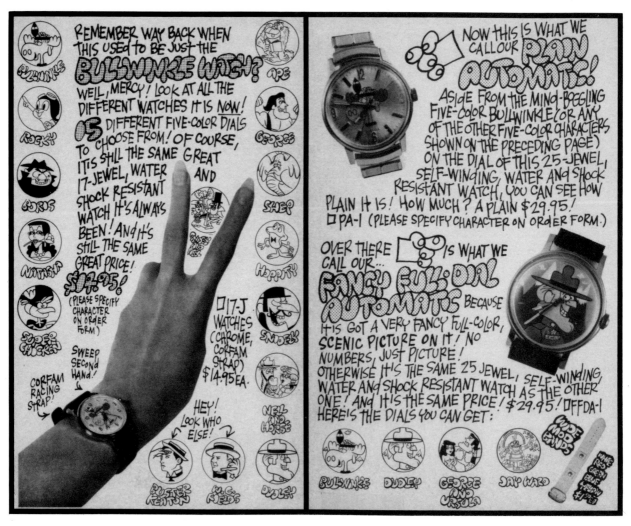

Bullwinkle and friends in another 1971 advertisement for Jay Ward watches.

Bullwinkle, 1971, Jay Ward. Hand-painted Bullwinkle, part of a series that included Dudley Doo Right, George and Ursula, and a caricature of Jay Ward. Size M.

Bullwinkle, made in 1971 for Jay Ward Productions. This is a seventeen-jewel movement watch and was part of a series of sixteen watches. Size M.

Bullwinkle, made in the mid-60s in Japan. Not for sale in the U.S. Size M.

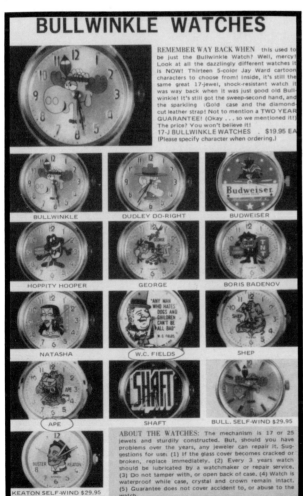

Bullwinkle and other Jay Ward character watches advertised in 1971.

Buster Brown, 1971 version of the original advertising watch. Size M.

Buster Brown, 1975 reissue. Size M.

Cat in the Hat, 1972, Lafayette. Watch came with a clear plastic, see-through back. Size M.

Casper, Nice Creation Co., mid-1960s. It was made in Japan, not for sale in the U.S. Size M.

Chip and Dale, 1970s, Japanese version. Not for sale in the U.S. This series of character watches, which also included Mickey Mouse and Donald Duck, came with solid backgrounds and was issued under the name Disney Time. Size M.

Chitty Chitty Bang Bang, 1971, Sheffield. Produced in conjunction with the movie. Size M.

Cool Cat, made by Sheraton in 1971. Part of a series of six. In this series the character's name is stamped on the back of the watch. Size M.

Cinderella, made in the 60s in Europe. Size M.

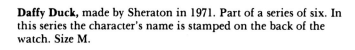

Cowboy, 1960, Ingraham. Optical animation. Size M.

Cowgirl, 1971. The famous artist, Joan Walsh, did the artwork for this watch. Size M.

Daffy Duck, made by Sheraton in 1971. Part of a series of six. In this series the character's name is stamped on the back of the watch. Size M.

Daffy Duck, made by Sheffield in 1971. This watch has large animated hands and is part of a series of four, including Porky Pig, Popeye, and Felix the Cat.

Dale Evans, made by Bradley in 1962. Size M.

Dale Evans, made by Ingraham, 1960. This is the Dale and Buttermilk reflector watch. This reflector series included two variations of Roy Rogers. Size M.

Daniel Boone, 1969, Swiss movement. Size M.

Davy Crockett, made in 1964 in England. It is the British version of our Davy Crockett. Size S.

Dennis the Menace, 1974, Bradley. Size M.

Donald Duck, 1964, Bayard. A series of clocks produced in France with limited distribution in the U.S. Others in the series include Pluto, Snow White, Mickey Mouse, Bambi, and Pinnochio. The heads of the characters wag back and forth. *Courtesy of Jack Feldman.*

Deputy Sheriff, 1966. One of the few watches made between 1958 and 1968. Size M.

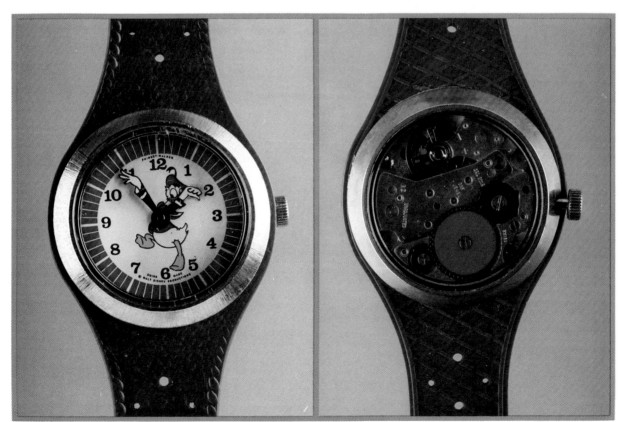

Donald Duck, Phinney Walker. This German company was most noted for their clocks. They only made two watches, the other being Mickey Mouse. The movement is shown. Size M.

Draemon the Cat is a Japanese superstar and was made in Japan in the mid-60s. Size M.

Dopey. Plastic version of Japanese character watches made in the 60s for sale out of the country. Size M.

Drooper, 1971, Hanna Barbera character. Size M.

Dudley Doo Right, made for Jay Ward in 1971. This is a seventeen-jewel movement watch and is part of a series of sixteen watches. Size M.

Elmer Fudd, made by Sheraton in 1971. Part of a series of six. In this series the character's name is stamped on the back of the watch. Size M.

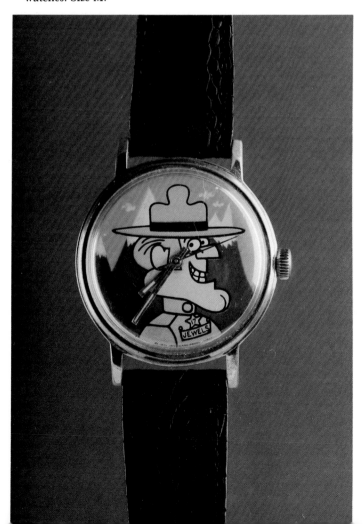

Dudley Doo Right, made for Jay Ward in 1971. This hand-painted dial has an automatic movement and was a part of a series of four that were all hand-painted. The other three in the series include Bullwinkle, George and Ursula, and a caricature of Jay Ward himself. Size M.

Felix the Cat, made by Sheffield in 1971. This features large animated hands and is part of a series of four that include Daffy Duck, Popeye, and Porky Pig.

Fred Flintstone, made by Prince Roable. Part of a series that included Pebbles and BamBam. Size M.

Flash Gordon, 1971, Precision Time. This is the only watch known to have been made by this company. Size M.

George of the Jungle, made for Jay Ward in 1971. This is a seventeen-jewel movement watch and is part of a series of sixteen watches. Size M.

GI Joe, 1966, Gilbert. Part of a series that included Batman, James Bond, and the Bronco. Size M.

Girl from Uncle, Bradley. A 1966 issue in conjunction with the television show. Size M.

Goofy, 1971 Helbros. Notice that the hands run backwards and one must tell time backwards. For this reason, the watch was not very popular and consequently limited numbers were made. This watch is the most valuable of its era. Size M.

Goofy, 1971, Helbros. The infamous backwards Goofy. By 1971 packaging had become of secondary importance to the manufacturers. Size M. *Courtesy of Jack Feldman.*

Hoppity Hoop, Made for Jay Ward in 1971. This seventeen-jewel movement watch was part of a series of sixteen watches. Size M.

Hot Wheels, Bradley, 1970. All three versions of Mattel's Hot Wheels series. Size M.

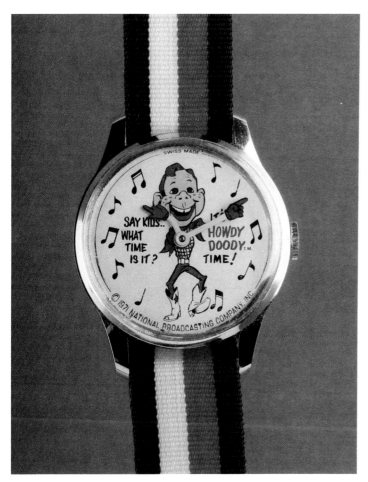

Howdy Doody, 1971, issued in conjunction with NBC. Size L.

Humphrey T. Bear, 1966, Japanese. Made for export. Size M.

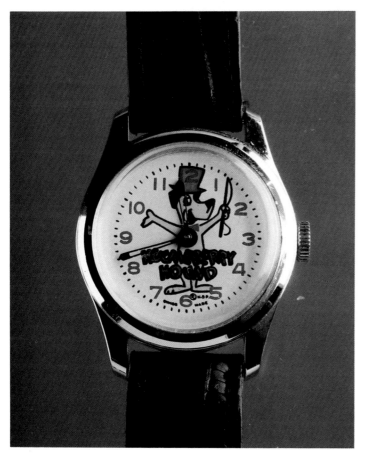

Huckleberry Hound, 1966, Bradley. One of the few watches made between 1958 and 1968. Size M.

Humpty Dumpty, 1967, foil-faced watch. Size M.

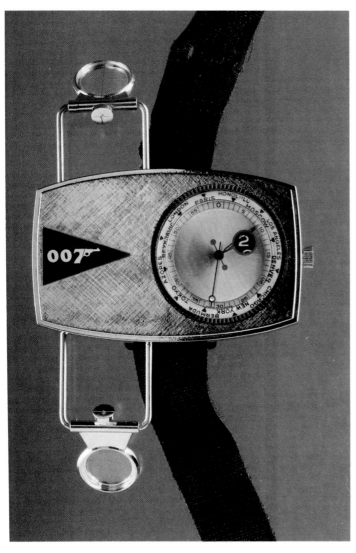

James Bond, 1966, Gilbert. Part of a series that included Batman, GI Joe, and the Bronco. Size M.

Lassie, 1960s, by Bradley in conjunction with the television show. Size M.

Jay Ward, 1971. Eleven of the sixteen watches in the series; Dudley Doo Right, Snidely Whiplash, Little Nell, Shep, George of the Jungle, Bullwinkle, Rocky, Natasha, Boris, Hoppity Hoop, and Super Chicken. Those missing are the ape from George of the Jungle, the hand-painted Bullwinkle, and Dudley Doo Right, George and Ursula, and the caricature of Jay Ward himself.

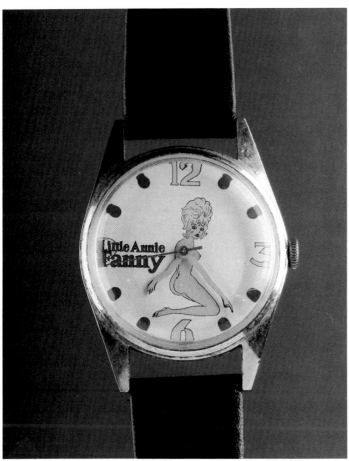

Little Annie Fanny, 1970. This watch commemorates the Playboy cartoon strip. Size M.

Little Nell, Made for Jay Ward in 1971. This is a seventeen-jewel movement watch and is part of a series of sixteen watches. Size M.

Little King. This watch was made in 1971 and is the only Little King watch made.

Little Red Riding Hood, 1960. On this watch the wolf is animated. Size S.

Louie, made in Hong Kong. This is a 60s watch and could be a prototype. Size M.

Man from Uncle, 1966, Bradley. This watch was made in conjunction with the television show. Size M.

Majorette, 1960, Ingraham reflector. Size M.

Merlin the Magic Mouse, made by Sheffield in 1971. After this watch, others were made with animated hands. Size L.

Mickey Mouse. A 1964 display for Bayard clocks. *Courtesy of Jack Feldman.*

Mickey Mouse, 1964, Bayard. A series of clocks produced in France with limited distribution in the U.S. Others in the series include Pluto, Snow White, Bambi, Donald Duck, and Pinnochio. The heads of the characters wag back and forth. *Courtesy of Jack Feldman.*

Mickey Mouse, 1966 version of Mickey Mouse. Unauthorized but unique since Mickey is sticking out his tongue. Size M.

Mickey Mouse, 1968, Timex. This is the watch that started the nostalgia craze a second time. Size M.

Mickey Mouse. The 1971 Helbros version of Mickey Mouse. The companion pieces are Minnie Mouse and the backwards Goofy. Size M.

Mickey Mouse, 1971. Made by Windert for sale only at the Disney parks. It was first made with white hands, but after 500 were made, the company realized their mistake and began producing them with yellow hands. Size M.

Mickey Mouse, 1971. Electric watches by Helbros, Elgin, and
Timex. Size M.

Mickey Mouse, 1972, Elgin quartz. Size M.

Mickey Mouse, 1972, Elgin quartz watches. Size M.

Mickey Mouse, 1971. Made in Israel, it has moving eyes. Size M.

Mickey Mouse. Mickey Mouse love watches from the 60s were made out of the country. Size M.

Mickey Mouse, 1972, Elgin quartz. Size M.

Minnie Mouse, made by Timex. These are the small and large versions of the 1968 Minnie Mouse Timex watch. The small one was issued with a statue similar to those statues issued in 1958. Some people consider the smaller version to have been issued prior to 1968, however Disney records prove that this watch was not licensed until 1968.

Minnie Mouse, 1971, Helbros. Companion pieces are Mickey Mouse and the backwards Goofy. Size M.

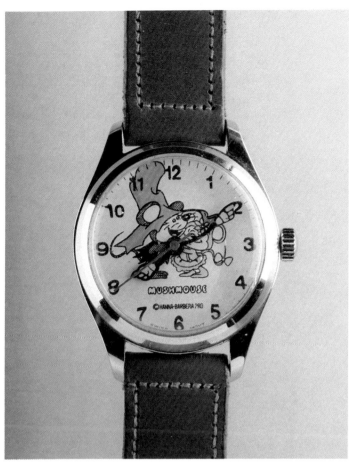

Mush Mouse, 1971, Hanna-Barbera, made in Japan. Not for sale in the U.S. Size M.

These watches were produced by U.S. Time in 1958 and were from a series of five, consisting of Mickey Mouse, Donald Duck, Alice in Wonderland, Cinderella, and Snow White. Size S.

Oliver, 1971, Sheffield. Made in conjunction with the movie. Size M.

Natasha, made for Jay Ward in 1971. This is a seventeen-jewel movement watch and is part of a series of sixteen watches. Size M.

Peace Mouse, 1970, by Peace Time Co. Size M.

Pebbles, made by Prince Roable, 1971. Part of the series that included BamBam and Fred Flintstone. Size S.

Piggy, 1972. It has a moving head. Size M.

Peter Pan, made by Rouan, 1972. This watch has a revolving disc. Part of a series that included Archie, Wendy, and Woody Woodpecker. Size M.

Pinnochio, 1964, Bayard. A series of clocks produced in France with limited distribution in the U.S. Others in the series include Pluto, Snow White, Mickey Mouse, Bambi, and Donald Duck. The heads of the characters wag back and forth. *Courtesy of Jack Feldman.*

Pinnochio, made by Bradley, 1966. Extremely rare. Size S.

Pluto, 1964, Bayard. A series of Disney clocks that was produced in France with limited distribution. The head of each character wags back and forth. The other clocks in the collection are Snow White, Mickey Mouse, Bambi, Donald Duck, and Pinnochio. *Courtesy of Jack Feldman.*

Popeye, and others in a series of Japanese watches including
Olive Oyl, Quick Draw McGraw, Tweety, and Tom and Jerry.
Size M.

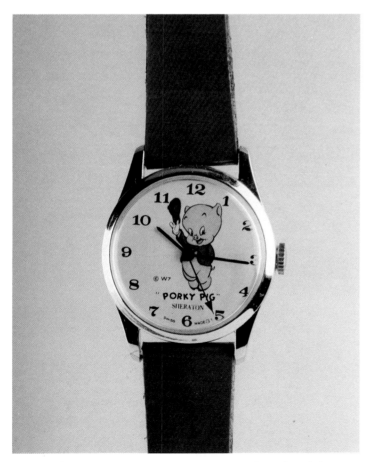

Porky Pig, made by Sheraton in 1971. Part of a series of six. In
this series the character's name is stamped on the back of the
watch. Size M.

Porky Pig, made by Sheffield in 1971. This large, animated-
hands watch is part of a series of four, including Felix the Cat,
Popeye, and Daffy Duck.

Puss-N-Boots, 1971, Japanese-made. Size M.

Quick Draw McGraw, made by Bradley, 1966. One of the few watches made between 1958 and 1968. Size S.

Quick Draw McGraw. This is part of a 1964 Japanese series of comic character watches which included Top Cat and Yogi Bear. Size M.

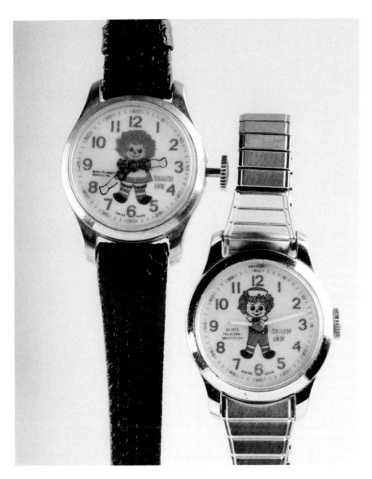

Raggedy Ann and **Raggedy Andy,** 1972, Bradley. Size M.

Roadrunner, 1971. It has a revolving disc with Wile E. Coyote chasing Roadrunner. Size M.

Roadrunner, made by Sheraton in 1971. Part of a series of six. In this series the character's name is stamped on the back of the watch. Size M.

Rocky, made for Jay Ward in 1971. This is a seventeen jewel movement watch and is part of a series of sixteen watches. Size M.

Roy Rogers, made by Bradley, 1962. This is a reissued, large version of the 1951 Roy Rogers.

Roy Rogers, 1960, Ingraham. This reflector watch features the back of the case engraved with the words "Many Happy Trails, Roy Rogers." Size M.

Roy Rogers, made by Bradley, 1962. This is a companion piece to Dale Evans. Size M.

Scooby Doo, 1970. This is the first Scooby watch made. Size M.

Scooby Doo, 1971. 21-jewel movement. This watch was a limited edition given to the production staff. Size M.

Sheffield, 1971. Merlin the Magic Mouse and the remaining four in the Sheffield series—Daffy Duck, Porky Pig, Felix the Cat, and Popeye.

Shep (the pink elephant from George of the Jungle), made for Jay Ward in 1971. This is a seventeen-jewel movement watch and is part of a series of sixteen watches. Size M.

Skipper, 1964. This is Barbie's friend. Size M.

Sheraton, 1971. Consisting of six watches in the series which include Daffy Duck, Cool Cat, Elmer Fudd, Wile E. Coyote, Porky Pig, and the Roadrunner.

Smokey Stover, made in 1971 by Hi-Time. Size L.

Smurf, 1960 model made in Germany. Instead of the familiar title "Smurf," it is called "Schlumpf." Size M.

Smokey the Bear, made by Hawthorne. The watch features hands that are two shovels and Smokey spelled out in a semicircle. This watch is often confused with the Smokey watch made by Bradley in 1973. In the Bradley model, the name "Smokey" is written horizontally. Size M.

Snoopy, Timex, 1969. One of the first watches made by Timex after the ten year hiatus between 1958 and 1968. Size M.

Snidely Whiplash, made for Jay Ward in 1971. This is a seventeen-jewel movement watch and is part of a series of sixteen watches. Size M.

Snoopy, 1969, Timex. The first one made. Size M.

Snoopy made by Timex. This watch not only came in these four variations, but also with a black background and a smaller watch with a powder blue background. It featured a silver—or gold-colored case. Size M.

Snoopy, 1969, Timex. Known as the Woodstock Snoopy. Size M.

Snoopy, made in the mid-70s in Japan. Not for sale in the U.S. Size M.

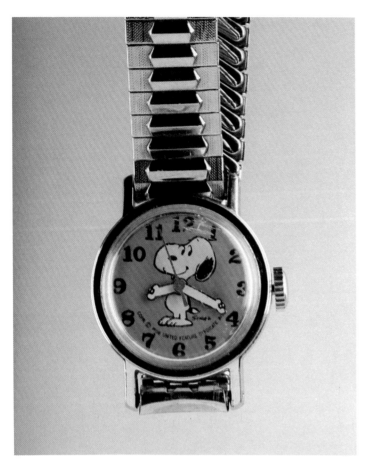

Snoopy, 1969, Timex. Part of a series of five different colors. Size S.

Space Mouse, 1965, made in Europe. Size M.

Snow White, 1964, Bayard. A series of Disney clocks produced in France with limited distribution in the U.S. Others in the series are Mickey Mouse, Pluto, Bambi, Donald Duck, and Pinnochio. All of the heads of the characters wag back and forth. *Courtesy of Jack Feldman.*

Spaceman, a 1960 watch made in Europe. Size M.

Spooky, 1966 prototype. Size S.

Superjew, 1972. Made in conjunction with the Yom Kippur War. Size M.

Super Chicken, made for Jay Ward in 1971. This is a seventeen-jewel movement watch and was part of a series of sixteen watches. Size M.

Superman, made by Bradley, 1962. One of the few watches made between 1958 and 1968. Size M.

Superman, 1968. The Superman rotating disc moves as the second hand moves. Size M.

Tammy Time Tell. Made in the 1960s to teach children to tell time. The companion piece is Terry Time Tell. Size M.

Tom and Jerry, 1966 version. Not available in the U.S. Size M.

Topo Gigio, 1970. A Japanese model to commemorate the mouse that starred on the Ed Sullivan Show. Size M.

Underdog, 1971, Leonardo. Size M.

Tweety, made by Time Setters, 1971. This is a companion piece to Sylvester. Size M.

Wendy, made by Rouan, 1972. The revolving disc watch which is part of the series that includes Archie, Woody Woodpecker, and Peter Pan. Size M.

Wile E. Coyote, made by Sheraton in 1971. Part of a series of six. In this series the character's name is stamped on the back of the watch. Size M.

Woody Woodpecker, made by Endora. It has a white plastic acrylic case with matching band and magnifying crystal. Size L.

Wizard of Oz, 1964. Made in conjunction to advertise the re-release of the movie. Size S.

Woody Woodpecker, 1972, Rouan. The revolving disc design is part of a series of watches that includes Wendy, Archie, and Peter Pan. Size M.

Woody Woodpecker, 1971. On this watch, Woody's friend is animated. Size M.

Yogi Bear, 1972, made by Prince Roable. Size M.

Yogi Bear, 1968, Bradley. One of the few watches made between 1958 and 1968. Size M.

Yosemite Sam, made by Time Setters, 1971. Size M.

FROM BUST TO BOOM: 1958-1972

WRISTWATCH

NAME	YR	NOTES
Alice in Wonderland	72	animated Madhatter
All-Star Baseball	66	black dial
	66	green dial
All-Star Football	66	autograph
Andy Panda	60s	blue plastic/Japanese
	71	red face
	72	small size
Ape--George		
of the Jungle	71	Jay Ward
Archie	72	Rouan/revolving disc
Astro Boy	60s	metal/Japanese
Ballerina	62	Bradley/hands at 1 and 10
	68	Bradley/hands at 4
BamBam	71	Prince Roable/small size
Barbie	63	facing left
	64	with Ken
	71	facing right/blue rim
Baseball Player	60	animated reflector
Batman	66	plastic by Gilbert
	71	Dirty Time
Boris	71	Jay Ward
Bozo the Clown	70	Capitol Record
Brick Bradford	71	Hi-Time
Bronco	66	Gilbert/plastic
Bronco Rider	60	animated reflector
Brutus	60s	metal/animated hands/Japanese
Buck Rogers	71	Huckleberry Time/pocket wristwatch
Buff and Jody	71	Sheffield/Time Tell
Bugs Bunny	72	animated hands/red bow-tie
	72	Bugs about baseball
	72	face and hand only/small size
	72	face only/small size
	72	Lafayette
Bullwinkle	71	Jay Ward
Buster Brown	71	Buster & Tige
Casper	60s	pink face
Cat in the Hat	72	see-through back
Chitty Chitty		
Bang Bang	71	Sheffield/from movie
Cinderella	60s	European
Cool Cat	71	Sheraton
Cowboy	69	moving gun
Cowgirl	71	Joan Walsh Anglund
Daffy Duck	71	Sheffield
	71	Sheraton
Dale Evans	60	flasher
	62	horseshoe/white
Daniel Boone	69	Powderhorn
Davy Crockett	64	English
Deputy Sheriff	60s	small size
Donald Duck	54	Swiss made
	60s	Phinney Walker
Dopey	60s	white clear plastic/Japanese
Draemon	70s	Japanese cat
Drooper	71	large size
Dudley Do-right	71	Jay Ward
	71	Jay Ward/hand-painted
Elmer Fudd	71	Sheraton
Felix the Cat	60s	Japan/black & white silhouette
	71	Sheffield
Flash Gordon	71	Precision Time
Fred Flintstone	71	Prince Roable/animated hands
G.I. Joe	66	Gilbert
George of the Jungle	71	Jay Ward

NAME	YR	NOTES
Girl from U.N.C.L.E.	66	pink dial
Goofy	71	backwards/Helbros
Hippie	70	Peace Time
Hopalong Cassidy	60	Great Britain/small/metal/ different collar
Hoppity Hoop	71	Jay Ward
Howdy Doody	71	animated hands
Huckleberry Hound	66	Bradley
Humphrey T. Bear	60s	Q and Q
Humpty Dumpty	67	blue metal face
James Bond 007	66	Gilbert
Jerry		
(of Tom and Jerry)	60s	white clear plastic/Japanese
Little Annie Fanny	70	Playboy bunny
Little King	71	O. Soglow
Little Nell	71	Jay Ward
Little Red		
Riding Hood	60s	animated wolf
Majorette	60	animated reflector
	69	Bradley
Man from U.N.C.L.E.	66	blue dial
Merlin the		
Magic Mouse	71	Sheffield
Mickey Mouse	60s	blue plastic/baseball player/ Japanese
	60s	moving eyes/made in Israel
	68	Timex
	69	Phinney Walker
	71	Elgin electric
	71	Helbros
	71	Helbros electric
	71	Helbros/day/date
	71	Timex electric
	71	Vantage/Disneyland/clear back
	71	Vantage/Disneyland/white hands
Minnie Mouse	71	Helbros
	71	Timex
Mush Mouse	60s	metal/animated hands/Japanese
Natasha	71	Jay Ward
Olive Oil	60s	metal/animated hands/Japanese
Oliver	71	Sheffield/from musical
Orphan Annie	71	Hi Time/large size
Peace Mouse	70	Peacetime Company
Pebbles	71	Prince Roable/small size
Peter Pan	72	Rouan/revolving disc
Pinnochio	60s	small size/on strings
	66	Bradley/lying on side
Popeye	60s	metal/animated hands/Japanese
	66	Bradley
	71	large, Sheffield/animated arms
Porky Pig	71	Sheffield
	71	Sheraton
Quick Draw McGraw	60s	green plastic/Japanese
	60s	metal/animated hands/Japanese
	66	Bradley
Raggedy Ann	71	small size
Roadrunner	71	revolving disc/with Wilee Coyote
	71	Sheffield
Rocky	71	Jay Ward
Rocky and Bullwinkle	60s	metal/Japanese
Roy Rogers	60	flasher
	60	reflector/running horse/ inscribed case
	62	white
	65	on Trigger
Scooby Doo	70	yellow-dial/animated hand

NAME	YR	NOTES
Shep	71	21-jewels/with date
	71	Jay Ward/pink elephant
Skipper	64	Mattel
Smokey Bear	71	Hawthorne
Smokey Stover	71	Hi Time
Smurfs--Schlumpf	60s	including Smurf band
Snidely Whiplash	71	Jay Ward
Snoopy	68	Timex/lying on top of doghouse
	69	black backgrnd/white Snoopy
	69	orange backgrnd/white Snoopy
	69	red backgrnd/white Snoopy
	69	yellow backgrnd/orange Snoopy
	69	yellow backgrnd/white Snoopy
Space Mouse	60s	blue face
Spaceman	67	yellow-gold face/Japan
Spooky	60s	blue and black
Super Chicken	71	Jay Ward
Superman	62	Bradley/Superman flying
	68	Revolving disc
Sylvester	71	Timesetters
Tammy Tell Time	60s	Teach Time
Terry Tell Time	60s	Teach Time
Tom and Jerry	60s	metal Tom w/fork/Jerry, second-hand/Japanese
Tom and Jerry	70	Jerry on second-hand
Top Cat	60s	red plastic/Japanese
	60s	white plastic/Japanese
Topo Gigio	70	big ears
Tweety	60s	metal/animated hands/Japanese
	71	Timesetters
Underdog	71	animated hands
Wendy	72	Rouan/revolving disc
Wilee Coyote	71	Sheraton
Wizard of Oz	60s	small size
Woody Woodpecker	71	moving Woodpecker
	72	plastic bubble
	72	Rouan/revolving disc
Yogi Bear	60s	with BooBoo/brown plastic
	68	Bradley
	71	Prince Roable
Yosemite Sam	71	Timesetters/animated hands

CLOCKS

NAME	YR	NOTES
Bambi	64	Bayard
Batman	69	Bradley
Beatles	71	yellow submarine
Bugs Bunny	60s	Seth Thomas
Donald Duck	64	Bayard
Lester Maddox	71	bat and chicken drum hands
Mickey Mouse	64	Bayard
	60s	green case/Phinney Walker
Pinnochio	64	Bayard
Pluto	64	Bayard
Popeye	68	Smits/animated Sweet-Pea
Sleeping Beauty	60s	round
Snow White	64	Bayard
Spiro Agnew	71	Peace hands

POCKET WATCHES

NAME	YR	NOTES
Buck Rogers	71	Huckleberry
Buster Brown	71	Huckleberry
Charlie Chaplin	71	Huckleberry
Flash Gordon	71	Huckleberry
Marilyn Monroe	71	Huckleberry
Mickey and Minnie	71	Love/Al Horen
Roy Rogers	59	Bradley/on Trigger
Superman	60	Bradley
Valentino	71	Huckleberry

Chapter Five
The Bradley Years:
1973-1985

At the end of 1972, Elgin Watch Company assigned the licensing they had acquired from Disney to their subsidiary, Bradley Watch Company. Elgin had acquired the license after Timex did not renew their contract. To this day, no one knows why Timex gave up this lucrative lease. Bradley embarked upon a new marketing theory. Instead of making the same watch for as many years as the public wanted, they chose to make many models of each character in each year. Between 1933 and 1972, there were approximately 380 different watches made. In the next 13 years, Bradley produced 1,800 different watch varieties, while other companies added approximately 200 different comic character watches to the marketplace. Since the Bradley watches dominated this period, I have chosen to call this time the "Bradley Years".

In mid-1991, the opportunity arose to purchase approximately 2,800 of these Bradley watches. What became important to me was the accompanying ad copy Not only did these watches include the Disney Bradley's, but they also included Star Trek, Star Wars, Muhammed Ali, and others.

As I studied this collection, it became apparent to me that Bradley had taken the same dial image and placed it in many case variations. The opposite was also true; they took a standard case and movement and changed the dials as they wished. Because of this practice, I decided that the case had to be significantly different from all the others in order to place it in my collection. One example of a distinct variation (and thus a different watch in my mind) is the lucite case which came in four shapes—round, square, rectangular, and octagonal. At the same time, Mickey came in a round metal case but with over 20 different dials—one with his foot out and one with his foot in. I chose one for my collection to represent this type of watch. After some pleasurable study, I have listed approximately 200 Bradley watches that I have included in my private collection. These are the watches shown in this chapter.

There were other significant watches made during the Bradley years. In 1968 Timex produced a series of four super heroes: Superman, Batman, Robin, and Wonder Woman. These were the first images we had of Robin and Wonder Woman. (Batman had been produced by Gilbert Watch in 1966, a plastic batwing without graphics, and Superman was made in 1939). In the same year, Dabs and Company issued their version of the super heroes: Superman, Batman, Wonder Woman, Spiderman, and the Hulk. These five watches have become some of the most copied watches in the comic character watch market. Full color ads for this series showed the watches in actual size. Those who could not afford the original cut out this actual size copy from the comic book and pasted the image on a watch face. However, there is a way to differentiate the two: the original had the words "Dabs and Company" by the number 11, the advertisement did not.

Animal, 1977, Picco. Seven-jewel movement. This is part of a series that included Fozzy Bear. Size M.

Batman, 1978, Timex. Part of a series that included Wonder Woman, Superman, and Robin. Size M.

Barbie, 1973, Bradley. The only plastic see-through Barbie that was ever made. Size M.

Batman, 1978, Dabs. Part of a series that included Superman, Wonder Woman, Spiderman, the Hulk, and the Joker. Size M.

In 1990, a 17-jewel Batman watch appeared at various antique shows which turned out to be a dial from new-old Timex stock. As the story goes, these 17-jewel watches were made to be sold in Canada, but because they had not sold by 1978, they were placed in warehouses where they were recently discovered. Except for the movement and screw-on case, this watch is identical to the original 1978 Timex. Each collector must decide for himself if he wishes to place a watch like this into his collection. I believe that many of these watches have attained a mystique of their own and belong in any comprehensive collection.

Some contemporary comic characters appeared for the first time during this period: Charlie Brown, Lucy, Nancy, and Sluggo. Some new Japanese cartoon characters such as Gundam and An Pan Man entered the marketplace.

Some of my personal favorites were made by Bradley during this period: (1) Mickey with a wagging head; (2) Minnie with a wagging head; (3) Mickey with Pluto, where Pluto's head wags; (4) Mickey and Minnie disco; (5) a series of Mickey sports such as basketball, jogging, baseball, football, tennis, and two soccers—one in which Mickey's foot moves as if he were kicking the ball; (6) a series of Goofy sports; (7) Mickey and Minnie playing tennis; and (8) Mickey and Minnie kissing. The animated Bradley watches, those featuring an object on the image that moves as the watch ticks, are extremely difficult to find for two reasons. They were produced in small quantities, and the movements were of such poor quality that people threw them away as they stopped working. Needless to say, not many have survived, making a collection of Bradley animated watches extremely difficult to acquire.

Bionic Woman, 1976, Bradley. Size M.

Big Jim, 1973, Bradley. Size M.

Boris and Natasha, 1984, by A & M. Size M.

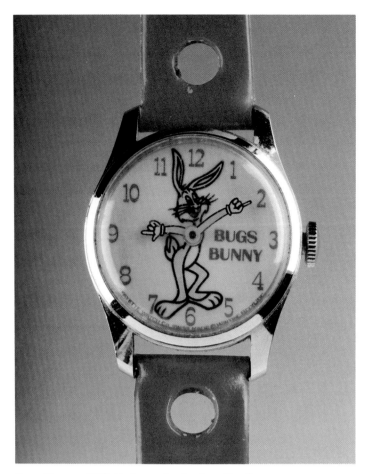

Bugs Bunny, 1974. Swiss made. Analog. Size S.

Care Bear, 1980 character from American Greeting cards. Size M.

Cabbage Patch Kids, 1983, Bradley. Size M.

Casper, 1974, Bradley. Size M.

Mickey watches made by Bradley may be divided into different categories making collecting fun and creative. Why not collect all Bradley Mickeys where he appears in long pants; Mickey tuxedo; Mickey in bluejeans, Mickey disco; and Mickey cowboy? Another category might be Mickey sports (listed earlier) and even Mickey commemoratives, of which there are three 50th Anniversary models.

The first one was produced in 1973 to commemorate the 50th Anniversary of the formation of the Disney company. It had embossed on the back, "50 Happy Years With Mickey". The second watch was released in 1978 to commemorate the 50th Anniversary of the first Mickey Mouse cartoon. Finally, in 1983 Bradley issued the 50th Anniversary of the Mickey Mouse watch. Even though these watches came in various sizes and shapes, there is a way in which you can tell them apart. The 1973 watch pictures Mickey's ears pointing to the number 12; the 1978 model shows his ears pointing to the number 11; and the 1983 image features his ears pointing to the number 1 symbolizing the anniversary of the 1st Mickey watch.

Bradley also produced a series of 17-jewel, 7-jewel, and high-fashion quartz watches, with primarily metal bracelets, which were meant to simulate the styles of Rolex and Cartier.

There were certain watches during this period that were distinctive enough to demand inclusion in any comprehensive collection. One pictures Mickey sticking out his tongue, and another shows Mickey with three hands. This obvious mistake features the hour and minute hands as animated Mickey hands, while a third hand is pictured separately. In addition to those I have mentioned, there were other watches made during these years that many collectors would find just as interesting and recommend that they, too, be included in any collection.

There is something for everyone in collecting comic character watches, and this chapter represents the fun of collecting the newer models. Because it would be so difficult to acquire them all, each person's collection will be different according to taste and stamina.

Cathy, Bradley, 1982. Size M.

Cat in the Hat, 1974, digital. The first one of its type. Size M.

Charlie Brown, 1974. This was the first time Charlie appeared on a watch. Size S.

Charlie Chaplin, 1985, Bradley. Backwards Charlie Chaplin that was never issued for sale to the public. Size M.

Charlie Chaplin. Part of the 1985 "Oldies" series made by Bradley that included W.C. Fields, the Three Stooges, Laurel and Hardy, Elvis Presley, Marilyn Monroe, Abbott and Costello, Emmett Kelly Jr. Just before going out of business, Bradley also produced a Charlie Chaplin watch that ran backwards. This watch was never issued for sale to the public, however it is available and sells for four times the value of any other watches in the series. The watches came in 2 sizes and 3 different bands.

Chipmunks, Bradley, 1984. Size S.

Cinderella, 1972, Bradley. Size M.

Cinderella, 1975, Bradley. Size M.

Cinderella, 1983, Bradley. Size M.

Cindy, 1973, animated moving slipper watch. Size M.

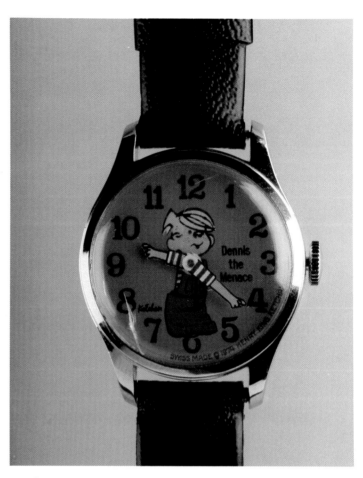

Dennis the Menace, 1974. Made exclusively for sale by J.C. Penny. Size M.

Daffy Duck, 1982, Hong Kong version. Size M.

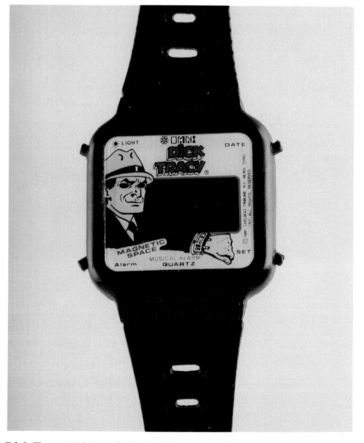

Dick Tracy, 1981, made by the Chicago Tribune to advertise the Dick Tracy cartoon adventure featuring him going to the moon. This watch is a two-tuned musical alarm watch. Size M.

6-PIECE "ANIMATED MOTION" 2-BELL ALARM ASSORTMENT

ASSORTMENT #9576HAAO

ASSORTMENT CONTAINS:

2 — 2089BWR8 — MICKEY MOUSE "Walking Action," red case
2 — 6485BBB8 — CARE BEARS™ "See-saw Action," blue case
2 — 6486BWR8 — STRAWBERRY SHORTCAKE™ "Rocking Basket Action," red case

CARTON: 5½ lbs.
9½" x 8" x 7"/.3 cu. ft.

MICKEY MOUSE ©Walt Disney Productions. CARE BEARS, STRAWBERRY SHORTCAKE ©American Greetings Corp.

12-PIECE QUARTZ 6" ROUND WALL CLOCK ASSORTMENT

ASSORTMENT #9580HADO

ASSORTMENT CONTAINS:

3 — 2100QWR8 — MICKEY MOUSE, animated hands, red case
2 — 2606QWB8 — DONALD DUCK, animated hands, blue case
3 — 6474QUU8 — STRAWBERRY SHORTCAKE,™ red case
4 — 6476QUU8 — CARE BEARS,™ blue case

MICKEY MOUSE, DONALD DUCK ©Walt Disney Productions. STRAWBERRY SHORTCAKE, CARE BEARS ©American Greetings Corp.

Bradley Time Division, Elgin National Industries, Inc.

CARTON: 10 lbs.
18" x 14" x 10"/1.45 cu. ft.

Character alarm and wall clocks, 1984.

Donald Duck. The back of the case of the Donald Duck 50th Anniversary watch.

Donald Duck, 1980. Made in Japan by Alba, a company known for full color dials. The companion pieces to this watch are Mickey Mouse, and Chip and Dale. Size M.

Donald Duck 50th Anniversary, made by Bradley. Size M.

Donald Duck 50th Anniversary, made by Bradley, 1984. This is a commemorative registered edition that came in three sizes. This large size with a black leather band, the medium size with a blue plastic band, and a small size with a black leather strap.

Always Timely... Always THE KING...

Fans of all ages can now have a great time with Elvis...listening to an official AM/FM clock radio... keeping time on-wrist or in-pocket...decorating with handsome wall clocks...all from Bradley Time...

E
L
V
I
S

6171QUU8 – An actual, colorful 12" record, ingeniously remade as a wall clock. The vibrant graphics will enhance any decor. Battery-operated quartz movement with sweep second.

5725PFW7 – A high-fashion wristwatch, sized for adults, with precision quartz movement. High-impact white plastic case and matching strap, black bezel ring.

5727F5W8 – Teens and adults will love this up-to-the-minute mechanical analog pocket watch, with stretchable telephone-type cord that hooks anywhere. Equally at home in pocket or purse. 2" diameter white case, clear cord.

6169QUU8 – A giant "wristwatch" wall clock, over 26" long. Quartz movement, battery-operated, with sweep second. Mock-gold case, black strap.

6170LUU8 – A portable, battery-operated AM/FM clock radio, featuring solid-state construction, direct tuning, great sound, dial light, snooze feature, headphone jack. Used as an alarm, one can wake to a buzzer or a favorite radio station. 4½" square x 4" deep.

BRADLEY TIME DIVISION/ELGIN NATIONAL INDUSTRIES, INC.
1115 Broadway, New York, N.Y. 10010

Elvis time pieces advertisement, 1984.

Felix the Cat, 1982, Hong Kong model. Size M.

Felix the Cat, 1982, Hong Kong model.

Flash Gordon, 1980, Bradley. One of only two Flash Gordons ever made. Size M.

Children's Character Watches

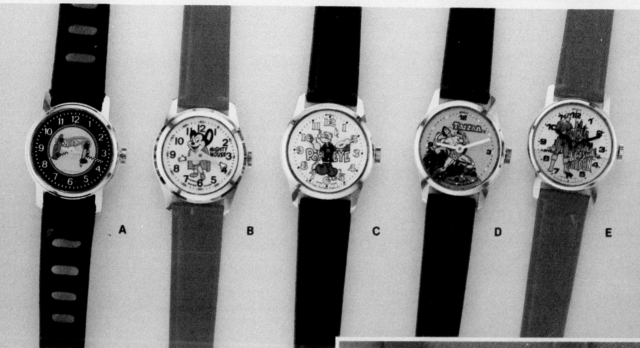

Action Heroes for Boys

A) **5400DJE2** — HOT WHEELS.® Two rotating race-cars, checkered hands. Chrome case, molded racing strap.

B) **5902DFR2** — MIGHTY MOUSE.® Chrome case, animated hands, 2-piece strap.

C) **5708DFE2** — POPEYE.® Spinach-loving favorite cartoon character. Chrome case, animated hands, 2-piece strap.

D) **5709DFE2** — TARZAN.® Jungle hero in full color. Chrome case, 2-piece strap.

E) **5742DFR2** — FLASH GORDON.® Space-age cartoon hero, soon to be seen again on film and TV. Chrome case, 2-piece strap, red sweep second.

Sesame Street Friends

F) **3070BFR2** — BIG BIRD.* Animated head and hands. Chrome case, 2-piece strap.

G) **3270BFF2** — COOKIE MONSTER.* Animated head and hands. Chrome case, 2-piece strap.

H) **3640BQF7** — BERT* & ERNIE.* — A special gift-pack that includes a FREE 7" Sesame Street "Birthday" record, and a party scene for kids to color. The colorful watch is small size, chrome case with red sweep second, one-piece fashion strap.

15

Children's Character Watches, 1980.

Flipper, 1973, Bradley. Size M.

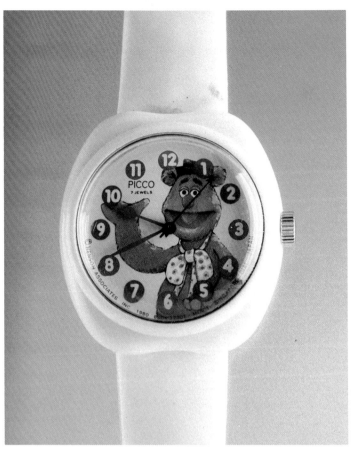

Fozzy Bear, 1977, Picco. Seventeen-jewel movement. This is part of a series that included Animal. Size M.

Fox and Hounds, 1980, Bradley. Size M.

Fozzy Bear, 1978, Picco. This is part of a series that included Kermit the Frog. Size M.

Fred Flintstone, 1978, Picco. Part of a series that included Pebbles and Bam Bam. Size M.

Garfield, 1978, Bradley. Size M.

Fred Flintstone, 1982, Bradley. Size M.

Gizmo, 1984, Nelsonic. Size M.

Goofy, 1982, Bradley. This is part of a series including Goofy playing soccer, tennis, and hockey. Size M.

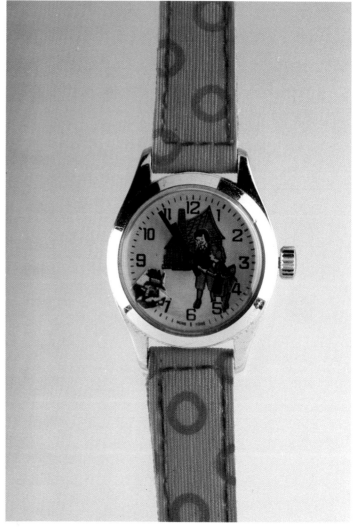

Gremlin, 1984. Size M.

Hansel and Gretel, 1974. The witch is animated. Size M.

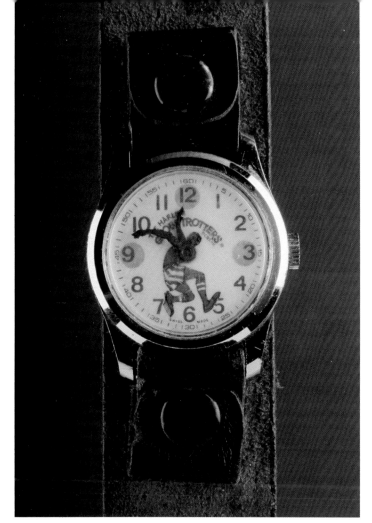

Harlem Globetrotters, 1975. Made in conjunction with CBS cartoon series. Size M.

Heathcliff, 1981, Bradley. Size M.

He-Man, 1985, Bradley. Size M.

Holly Hobby, 1980, Bradley. Size M.

Inspector Gadget, 1984, Bradley. Size S.

The Hulk, 1978, Dabs. Part of a series that included Batman, Superman, Wonderwoman, the Joker, and Spiderman. Size M.

The Joker, 1978, Dabs. Part of the series that included Batman, Superman, Wonder Woman, the Hulk and Spiderman. Size M.

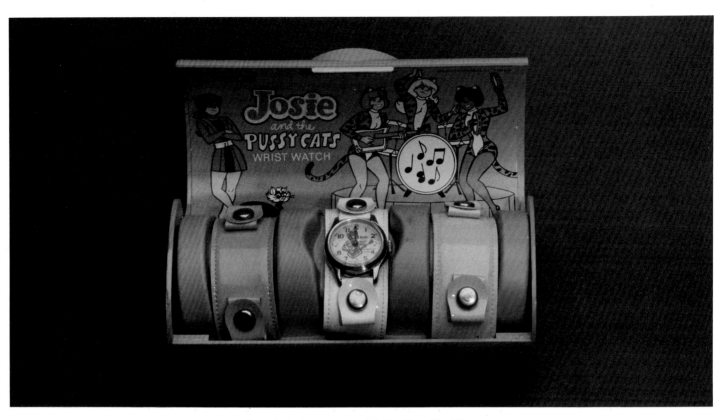

Josie and the Pussycats, 1974, Bradley. Size M.

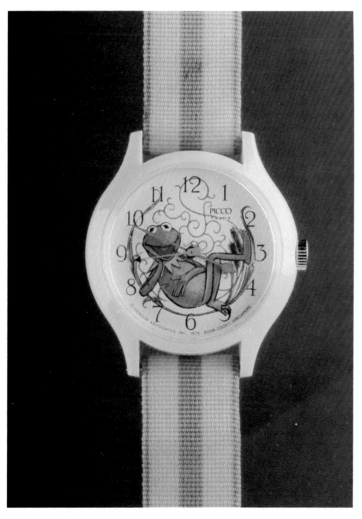

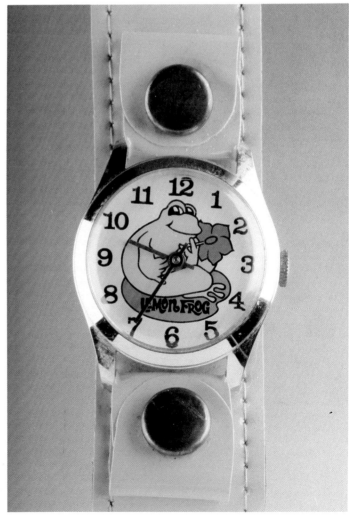

Kermit the Frog, 1978, Picco. Part of a series that included Fozzy Bear. Size M.

Lemon Frog, 1980. Size M.

Lone Ranger, Bradley, 1980. This is a reissue for the television show. Size M.

Lucy, 1974, Times. Size S.

Mary Poppins, 1965. Made in conjunction with the movie. Size M.

Merrie Mouse, Bradley, 1978 model. Size M.

Mickey and Minnie. These are the only Mickey and Minnie watches made during this period. One with Mickey and Minnie playing tennis comes with a sweep second hand which is a tennis ball. The Mickey and Minnie Disco is unusual in that both heads move and the outer ring rotates. This watch was part of a series that included Minnie Mouse moving head, Mickey Mouse moving head, and an animated Pluto. Tennis watches Size M, others S.

Mickey and Pluto, made by Bradley. Part of the series that included the Mickey moving head, the Minnie moving head, and the Mickey and Minnie Disco. Pluto was animated and is extremely valuable because only a limited quantity was made. Size S.

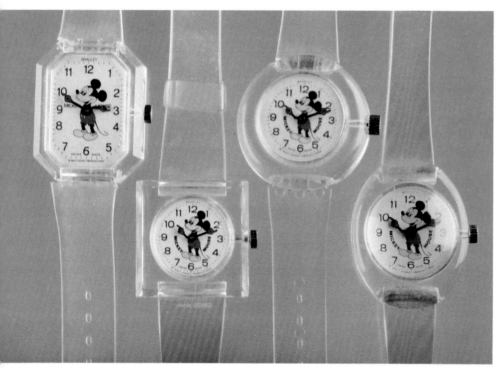

Mickey Mouse, made by Bradley. These are Lucite, clear plastic, see-through watches which came in these four shapes. They are serviced by entering the watch from the front.

Mickey Mouse, made by Bradley. Considered extremely collectible because of the obvious mistake in the design. Not only are there two animated hands, but notice the third hand that Mickey has resting on his hip. Size S.

Mickey Mouse Jogger, made by Bradley. Mickey is pictured on a celluloid disc. The disc rotates and it appears that Mickey is jogging around the watch. Size M.

Mickey Mouse, made by Bradley. In this model Mickey's head bobs back and forth. This is part of a series which included a Minnie moving head, Mickey and Pluto, and Mickey and Minnie Disco. Size S.

Mickey Mouse, made by Bradley. One of the many different 17-jewel fashion watches made by Bradley. Size M.

Mickey Mouse Auto Racer, made by Bradley, with a black acrylic case. Size M.

Mickey Mouse, made by Bradley. Black plastic acrylic Mickey Mouse. It also came in white plastic. Size M.

Mickey Mouse, On the left is the Anniversary watch, 1973. This was made to commemorate the 50th anniversary of the formation of the Walt Disney company. Inthe center are two versions of the 1978 50th anniversary of the first Mickey Mouse cartoon. A version of the 1983 50th anniversary of the Mickey Mouse watch is on the right. Size M.

Mickey Mouse, 1970s, Bradley quartz alarm. Size M.

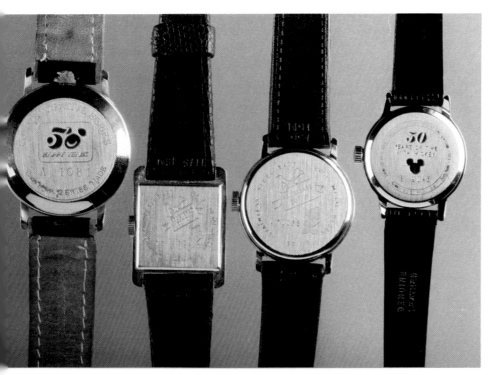

Mickey Mouse Backs of the Anniversary watches.

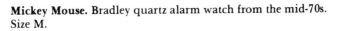

Mickey Mouse. Bradley quartz alarm watch from the mid-70s. Size M.

Mickey Mouse. Commemorative watch for the only team championship golf tournament that Disney held at Walt Disney World in 1974. After this tournament the format was changed and commemorative watches have been given each year. Size M.

Mickey Mouse, 1973, Bradley Mickey Sportsman. Size M.

Mickey Mouse, Bradley, 1975, seventeen-jewel tank watch. Size M.

Mickey Mouse, 1975, Bradley seventeen—and seven-jewel movement watches. Size M.

Mickey Mouse. 1976 commemorative for the bicentennial. Size M.

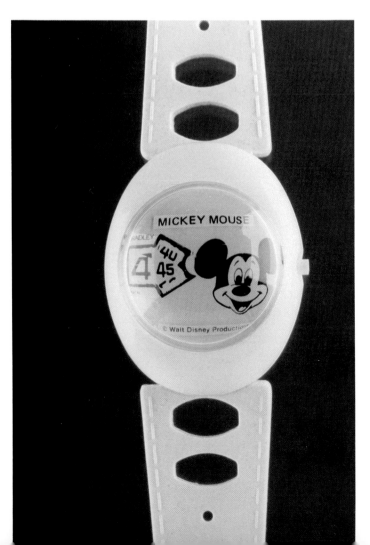

Mickey Mouse, Bradley prototype, 1975. Size M.

Mickey Mouse, 1978, Bradley Quartz alarm chronograph. Size M.

Mickey Mouse,' 1979, Bradley, Motorcycle Helmet. One of seven watches in which Mickey is in long pants. Size M.

Mickey Mouse, 1978. A Bradley watch showing Mickey playing basketball, tennis, football, and baseball. Size M.

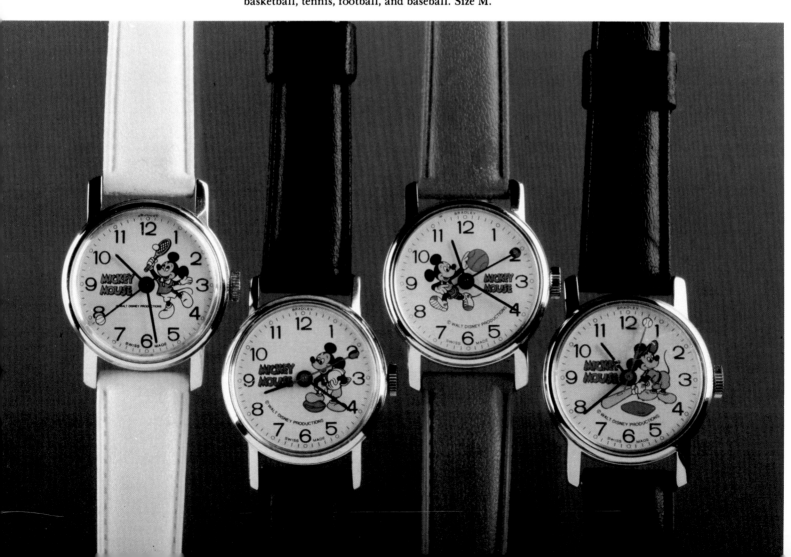

Mickey Mouse, 1978, Bradley Sportsman. Size M.

Mickey Mouse, 1978, Bradley Mickey Soccer watches. Both have Mickey's foot animated. Size M.

Mickey Mouse, 1976, Bradley digital. One of only two models made in this fashion. Size M.

Mickey Mouse, 1980, Bradley designer series. Size M.

Mickey Mouse, 1980, Bradley Mickey Disco. One of only seven Mickey Mouse watches where he appears in long pants. Size M.

Mickey Mouse, 1980, Bradley designer series of watches. Size M.

Mickey Mouse, 1980, Bradley designer series. Size M.

Mickey Mouse, 1980, Bradley designer series. Size M.

Mickey Mouse, 1980, Bradley seven-jewel movement watch. Size M.

Why should kids have all the fun?

ALL QUARTZ…ALL GORGEOUS

MQ6

1462TBE6—All silvertone; vintage Mickey with animated hands; sized for ladies.
1463TBE6—Same, all goldtone.

The Supreme Mickey Mouse Watch in 18K Gold

1293SBD6—Golden Coin for Ladies. 18K gold case. Raised markers; raised and engraved Mickey with animated hands; brown leather strap.
1294TBD6—Golden Coin for Men. Same as above.

1473SBE6—Ladies' Goldtone Coin. Charcoal dial with raised golden markers. Raised and engraved goldtone Mickey with animated hands.
1474TBE6—Same, for men.

1477SBE6—Goldtone wide-bezel case with engraved numerals; golden Mickey with animated hands against rich black dial.
1478TBE6—Same, for men.

1475SBE6—Bold silvertone case with goldtone rivet markers and bezel ring; calendar; golden Mickey with animated hands against rich black dial; durable black rubber strap.
1476TBE6—Same, for men.

1485TBE6—Men's High-Tech Cushion. Black anodized case, with matching strap. Full-figure goldtone Mickey with animated hands; calendar.
1486SBE6—Same, for ladies, without calendar.

BRADLEY

© Walt Disney Productions

Mickey Mouse character watches advertisement, 1983.

Mickey Mouse, 1980, Bradley Designer series. Size M.

Mickey Mouse Gold Coin watch, made by Bradley. The fluted bezel and carved relief dial makes this watch different from all other watches. Size M.

Mickey Mouse, 1982, Bradley. Called "Mickey Western." One of seven watches on which Mickey is in long pants. Size M.

Mickey Mouse. 1982, Bradley silver coin. Size M.

Mickey Mouse, 1983, Bradley collector series. Size M.

Mickey Mouse. Bradley 50th Anniversary of the Mickey Mouse watch, 1983. Two versions, one with and one without a calender. Size M.

Mickey Mouse, 1983, Bradley collector series. Size M.

Mickey Mouse and friends, 1973.

Mickey Mouse, made by Bradley. One of the 12 of the 1984 designer series. Size M.

Mickey Mouse, Bradley. Oval-shaped, 17-jewel movement. One of the series of over twelve 17-jewel movement watches made in different sizes and shapes.

Mickey Mouse, made by Bradley. Part of the designer series put out in 1984. This is one of twelve in the series and all have become extremely collectible. Size M.

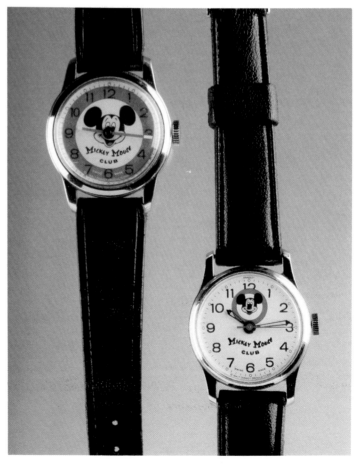

Mickey Mouse Club, Bradley, 1978. Size S.

Mighty Mouse, Bradley, 1981. Size M.

Mighty Mouse, 1979, Bradley. Size M.

Mighty Mouse, 1981, Bradley. Size M.

Mighty Mouse, 1980, Bradley. Size M.

Minnie Mouse Moving Head, made by Bradley. One of the series that included the Mickey moving head, the Mickey and Pluto, and the Mickey and Minnie Disco. There were less than 350 of these made and they are extremely rare. Size S.

Mork, 1978, Bradley. Size M.

Ms. Pacman, 1980, Bradley.

Nancy, made by Bradley, 1974. This watch is part of a series that included Sluggo and should have animated hands. Size S.

Muhammad Ali, 1980, Bradley. One of only two Ali watches made. Size M.

New Zoo Review, 1975, consisting of Rhino, Owl, and Frog. Size S.

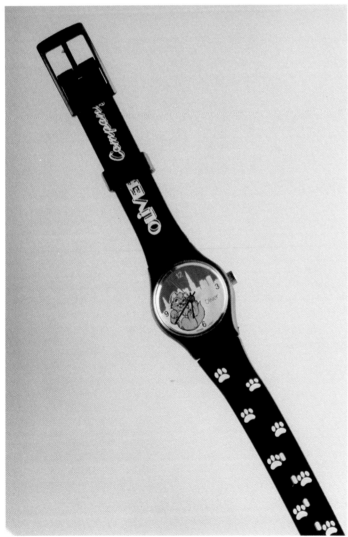

Oliver, 1980s promotional for a cartoon series. Size M.

Orphan Annie, 1975. Made in conjunction with the New York Daily News. Size M.

Orphan Annie, 1980, Clinton watch. Size M.

Pebbles and Bam Bam, 1978, Picco. Part of a series that included Fred Flintstone. Size M.

Pacman, 1980, Bradley. Size M.

Peter Rabbit, 1980, Bradley. Size M.

Penny, 1978, Bradley. Size M.

Pink Panther, 1979 model made in Hong Kong. Size M.

Pinnochio, 1974, Webster. Jiminy Cricket is animated. Size M.

Popeye, made by Bradley, 1974. One of the prettiest of the Popeye watches made. Size M.

Pluto, 1980. This is a Hong Kong version of Pluto. Size M.

Puss-N-Boots, 1970. Size M.

Robin, 1978, Timex. Part of the series that included Batman, Superman, and Wonder Woman. Size M.

Robin Hood, 1974, Bradley. Size M.

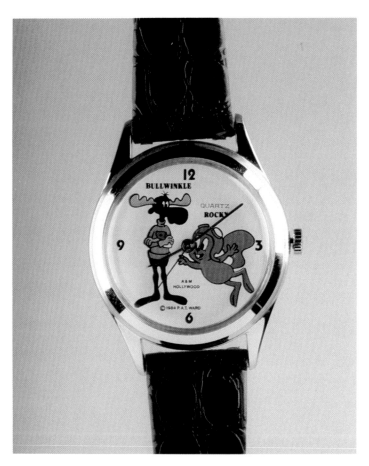

Rocky and Bullwinkle, 1984, by A & M. Size M.

Sam the Eagle. Made as a commemorative of the 1984 Olympics in Los Angeles. Size M.

Roy Rogers & Dale Evans, 1985, Bradley. Size M.

Bert and Ernie, 1977, Bradley. Size M.

12-PIECE TALKING ALARM ASS'TMENT #9617HADO
Wake Up to the Voices of Favorite Characters

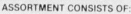

ASSORTMENT CONSISTS OF:
3 — 4081NBUU —BIG BIRD*
3 — 4055NBUU —MUPPET* SCHOOLHOUSE
3 — 6287NBUU —STAR WARS*
3 — 6293NBUU —SMURFS™

© Muppets, Inc. *MUPPET, Big Bird are trademarks of Muppets, Inc. STAR WARS
© Lucasfilm Ltd. SMURFS © Peyo, lic. by Wallace Berrie & Co, Inc.

PRODUCT OF
BRADLEY TIME
A DIVISION OF ELGIN NATIONAL IND INC

BRADLEY TIME DIVISION/ELGIN NATIONAL INDUSTRIES, INC.
1115 Broadway, New York, N.Y. 10010

SHIPPING WT: 25 lbs.
24" x 15" x 19"/3.96 cu. ft.

Smurf and other character clocks advertisement, 1984.

Oscar the Grouch, 1977, Bradley. Size M.

The Count, 1977, Bradley. Size M.

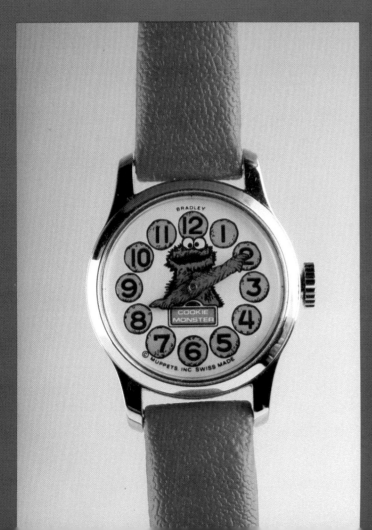

Cookie Monster, 1977, Bradley. This watch also came with a moving head as did Big Bird. Size M.

Big Bird, 1977, Bradley. This watch also came with a moving head, as did the Cookie Monster. Size M.

Six Million Dollar Man, 1976, Bradley. Size M.

Ernie, 1977, Bradley. Size M.

Skippy, 1980. The eyes on this watch move back and forth. Size M.

Sluggo, Bradley, 1981. This watch should have animated hands. Size S.

Smurf, 1983, Bradley. Size M.

Smokey the Bear, 1979, Bradley. Size M.

Snoopy, 1976, Timex. Called the "Denim Snoopy." Size M.

Snoopy, 1979, Timex. Called the "Flying Ace Snoopy." Size M.

Snoopy, 1977, Timex. Called the "Tennis Snoopy." Size M.

Spiderman, 1978, Dabs. Part of the series that included Batman, Superman, Wonder Woman, the Joker, and the Hulk. Size M.

Sport Billy, 1982, Bradley. The foot on this watch is animated. Size M.

Star Wars, 1977, German models of Star Wars—"Kreig Sterne." Obviously more valuable than the American models because of the German writing. Size M.

Star Wars, 1980, Bradley. Size M.

12-PIECE SCENI-CLOCK ASSORTMENT #9618HADO

Colorful Standing Clocks to Decorate any Kid's Room

STAR WARS

MAY THE FORCE BE WITH YOU!

Strawberry Shortcake

THE SESAME STREET PLAYERS

ASSORTMENT CONSISTS OF:
2 — 2065QUU8 — DISNEY CASTLE
2 — 4900QUU8 — C.T.W. MUPPET* PLAYHOUSE
2 — 6290QUU8 — HOLLY HOBBIE* HOUSE
3 — 6291QUU8 — STRAWBERRY SHORTCAKE* BERRY
3 — 6292QUU8 — STAR WARS*

Walt Disney Productions. * Muppets, Inc. HOLLY HOBBIE ©1972 American Greetings Corp.
STRAWBERRY SHORTCAKE © 1980 American Greetings Corp. STAR WARS Characters ©Lucasfilm Ltd.

PRODUCT OF
BRADLEY TIME
A DIVISION OF ELGIN NATIONAL IND INC

BRADLEY TIME DIVISION/ELGIN NATIONAL INDUSTRIES, INC.
1115 Broadway, New York, N.Y. 10010

SHIPPING WT: 17 lbs.
12" x 16" x 28"/3.11 cu. ft.

Advertisement for character clocks, 1984.

AM/FM CLOCK/RADIO ASSORTMENTS

Fully Battery-Powered Solid-State Radios with Quartz Analog Clocks featuring Direct Tuning, Music or Alarm, Dial Light, Snooze Feature

In Sturdy, Stackable, See-Thru Packages

12-PIECE ASSORTMENT #9577HAD0

2 – 2087LUU8 – MICKEY MOUSE, red and white
2 – 6486LUU8 – STRAWBERRY SHORTCAKE,™
 red and white
1 – 6469LUU8 – STAR WARS,® black and silver
3 – 6482LUU8 – CARE BEARS,™ blue and white
2 – 6483LUU8 – CHIPMUNKS,™ red and white
2 – 6493LUU8 – ROBO-FORCE,™' black and silver

CARTON: 18 lbs.
20″ x 16″ x 11″/2.03 cu. ft.

6-PIECE ASSORTMENT #9585HAA0

2 – 6482LUU8 – CARE BEARS,™ blue and white
1 – 6468LUU8 – STRAWBERRY SHORTCAKE,™
 red and white
2 – 2087LUU8 – MICKEY MOUSE, red and white
1 – 6493LUU8 – ROBO-FORCE,™ black and silver

CARTON: 8 lbs.
19½″ x 6″ x 10½″/.71 cu ft.

Bradley Time Division, Elgin National Industries, Inc.

Advertisement for character clocks, 1984.

Star Wars, 1980, Bradley. Size M.

Strawberry Shortcake, 1980, Bradley. Size M.

Superman, 1978, Timex. Part of a series that included Batman, Robin, and Wonder Woman. Size M.

Superman, 1978, Dabs. Part of a series that included Batman, Wonder Woman, the Joker, the Hulk, and Spiderman. Size M.

Tarzan, 1983, both models by Bradley. Size M.

Superman, 50th Anniversary watch made as a promotional for Armour Meat Company. Size M.

Tom and Jerry, 1985, Bradley. Size M.

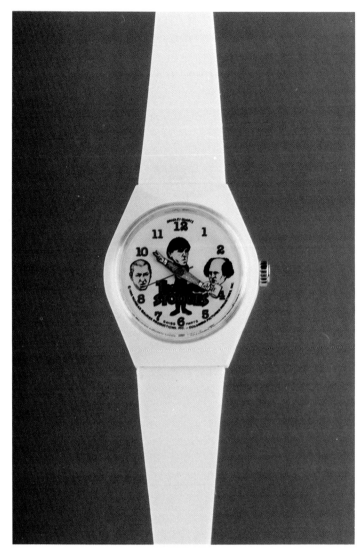

Three Stooges, 1985, "Oldies" series by Bradley that included W.C. Fields, Charlie Chaplin, Laurel and Hardy, Elvis Presley, Marilyn Monroe, Abbott and Costello, and Emmett Kelly Jr. The watches came in two sizes and three different bands.

Touche Turtle. 1980 watch made in Hong Kong. Most watches made by this manufacturer have this type of coloration. Size M.

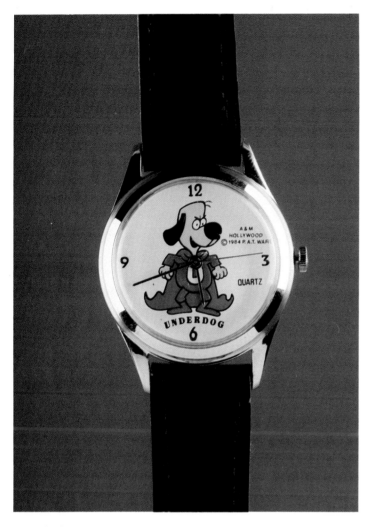

Underdog, 1984, put out by A & M. Size M.

Winnie the Pooh, 1977, seven-jewel movement. Made for sale at Sears. Size S.

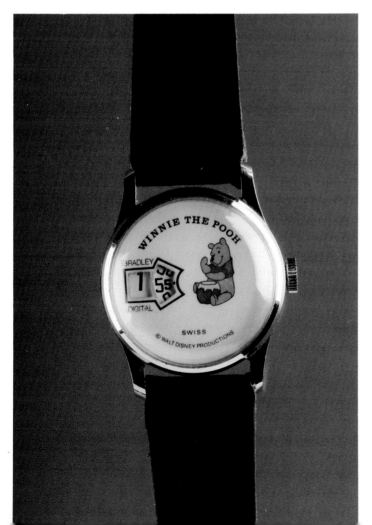

Winnie the Pooh, 1977, Bradley. One of the few digital comic character watches made. Size M.

Winnie the Pooh, 1985, Bradley. Size M.

Wonder Woman, 1978, Dabs. Part of a series that included Batman, Superman, the Joker, the Hulk, and Spiderman. Size M.

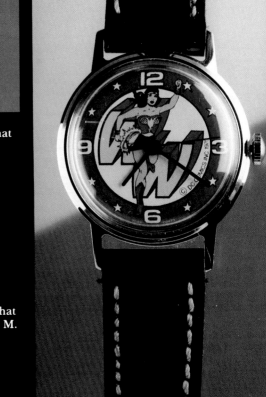

Wonder Woman, 1978, Timex. Part of series that included Batman, Superman, and Robin. Size M.

Woodsey Owl, 1975. Smokey the Bear's friend. Size M.

Woody Woodpecker, 1982, Hong Kong model. Size M.

Wuzzies, 1985, Bradley. Size M.

THE BRADLEY YEARS: 1973-1985

WRISTWATCHES

NAME	YR	NOTES
Animal	77	Picco/orange plastic
Aquaman	78	DC Comics/with date
Barbie	73	animated hands/plastic, see-thru case
Batman	78	Dabs
	78	Timex
Bert and Emie	80s	Bradley/Sesame Street
Betty Boop	85	hearts and moving eyes
Big Bird	80s	Bradley/Sesame Street/moving head
Big Jim	73	Bradley
Bionic Woman	76	red
Boris and Natasha	84	quartz
Brother Bear	84	Armitron/blue plastic
Bugs Bunny	76	Great America for Marriott Corp
Buster Brown	70s	Buster and Tige
Cabbage Patch Kids	83	small/green
Casper	74	Bradley/blue face
Cat in the Hat	74	blue digital
Cathy	82	Bradley
Charlie Brown	74	yellow-dial/animated hands
Charlie Chaplin	80s	Japanese/moving hands
Chip and Dale	80	green disc
Chipmunks	84	small size
Cinderella	73	Bradley/pink circle
	73	Bradley/with prince
	75	Bradley/blue dress/running
Cindy	73	moving slipper
Commando	66	Bradley
Cookie Monster	70s	Bradley/moving head
	80s	Bradley/Sesame Street
Count	80s	Bradley
Curious George	80s	yellow plastic/small size
Daffy Duck	80s	Hong Kong
Dennis the Menace	74	animated hands/orange dial
	74	Bradley/with dog
Dick Tracy	80s	lunar lander/Chicago Tribune
	81	Chicago Tribune/paper car/watch inside
Donald Duck	80	Alba/green face
	84	50th anniversary/large/gold
	84	Bradley/flashing quartz
	84	Happy Birthday, Donald/LCD
E.T.	82	face
Elliott the Dragon	77	Bradley
Emie	80s	Bradley/Sesame Street
Fazi Bear	77	Picco/red plastic
	77	Picco/white plastic
Felix the Cat	80s	Hong Kong
	80s	yellow dial
Flash Gordon	80	Bradley
Flipper	73	Bradley
Fox and Hounds	80	Bradley/small size
Fred & Pebbles	80s	Bradley
Fred and Dino	78	Picco/white plastic
Funky Fanthom	73	Hanna Barbera
Garfield	78	on the moon
Gizmo	84	small size
Golden Girl	70s	Bradley
Goofy	75	Bradley/small size
	82	Bradley/hockey
	82	Bradley/soccer
	82	Bradley/tennis
Gremlins	84	green face
Gumby	80s	green/Hong Kong
Hagar	80s	Sutton Time/cartoon strip
Hansel and Gretel	74	animated
He-Man	80s	Bradley
Heathcliff	81	Bradley
Herbie Goes Bananas	80s	from movie
Hi and Lois	80s	Sutton Time/cartoon strip
Holly Hobbie	78	square
Howard the Duck	80s	issued for movie
Hulk	78	Dabs/small size
Inspector Gadget	84	Bradley
Joker	78	Dabs
Josie and the Pussycats	74	Bradley
Kermit the Frog	77	Picco/yellow plastic
Leo the Lion	70s	Happy Hour Creations
Lone Ranger	80	Bradley
Louie Duck	80	Hong Kong/round
Lucy	74	Letters under feet
Marvin	83	Bradley
Mary Poppins	70s	Bradley/from movie

NAME	YR	NOTES
Merrie Mouse	70s	animated hands
Mickey and Minnie	78	Bradley/Mickey & Minnie love
	80	Bradley/disco/rotating disc
Mickey and Minnie Tennis	78	Bradley
Mickey and Pluto	83	Bradley/animated Pluto head
Mickey Love	70s	flower
	70s	love balloon
	70s	moving flower
Mickey Mouse	70s	tongue out
	70s	Bradley/quartz
	70s	Elgin/quartz
	73	50 years w/Mickey/ears at 12
	73	Bradley/sportsman/calendar
	73	Time Teacher
	74	golf classic
	76	Bradley/Bicentennial
	76	Bradley/digital
	76	Bradley/Three Hands
	76	white plastic/acrylic/oval
	76	white plastic/acrylic/rectangular
	76	white plastic/acrylic/round
	76	white plastic/acrylic/square
	78	50th birthday/round/ears at ll
	78	50th birthday/square/ears at 11
	78	Bradley/radio watch
	78	Bradley/auto racer
	78	Bradley/baseball
	78	Bradley/basketball
	78	Bradley/football
	78	Bradley/moving head
	78	Bradley/small
	78	Bradley/soccer/moving foot/grn
	78	Bradley/soccer/moving foot/wht
	78	Bradley/sportsman/rotating bezel
	78	Bradley/tennis
	78	white/plastic/digital
	79	Bradley/red racing helmet
	80s	Bulova accutron
	80	Bradley/gold/circular case
	80	Bradley/Mickey rotates/jogger
	80	Bradley/scalloped edge/gold coin
	80	Bradley/silver coin
	80	Bradley/tuxedo
	80	Bradley/white face/gold scalloped edge
	80	disco
	82	17-jewel/rectangular
	82	Bradley/cowboy
	82	Bradley/rotating bezel
	82	Ingot
	83	50 Years of Time/calender
	83	50th anniversary of watch/gold/ears point to 1
	83	Bradley/alarm/chronograph
	83	Bradley/gold-tone/stripped bezel dial
	83	Bradley/gold/hexagon
	83	Bradley/silver-tone/gold-tone/rivet markers & bezel
	83	Bradley/silver/hexagon
	84	Bradley/black square/black anodized case & band/gold Mickey
	84	Bradley/flashing quartz
	84	Bradley/flip-top/5-function/LCD
	84	Bradley/gold-tone case/black dial/Roman num on bezel
	84	Bradley/gold-tone/charcoal dial/Mickey gold
	84	Bradley/gold-tone/gold face/Mickey walking
	84	Bradley/round/silver/rivets
	84	Mickey face/gold bar band
Mickey Mouse Club	76	Bradley/large face
	76	Bradley/large face/white no.'s
	78	Bradley/small face
	78	small face/see-through letters
Mighty Mouse	70s	Bradley/flying
	79	Bradley/animated hands
	81	Bradley
	85	Bradley/red plastic
Million Dollar Man	76	picture
Minnie Mouse	70s	fat/Bradley
	78	Bradley/moving head

NAME	YR	NOTES
Miss Piggy	84	Bradley/flashing quartz
	80	Picco/purple face
Mork	78	picture
Mouse Club	82	Disneyana Collector's Club
Ms. Pacman	80	Bradley
Nancy	74	Bradley
	80s	Sutton Time/cartoon strip
New Zoo Revue--Elephant	75	face
New Zoo Revue--Frog	75	face
New Zoo Revue--Owl	75	face
Orphan Annie	75	New York News
Oscar the Grouch	80s	Bradley/Sesame Street
Pebbles and BamBam	78	Picco/white plastic
Penny	78	Bradley
Peter Pan	80s	Hong Kong
Peter Rabbit	80	Bradley
Piggy	73	moving head
Pink Panther	79	large size
Pinnochio	74	moving Jiminy Crickett
Pipi Longstocking	70s	Bradley
Pluto	80s	Hong Kong
Polly Pal	74	Leon Jason
Popeye	74	Bradley/animated hands
Puss 'n Boots	70s	knife & sword/small size
	70s	large Puss
Raggedy Andy	73	small size
Raggedy Ann and Andy	75	Seesaw Hearts
Robin	78	Timex
Robin Hood	74	Bradley/character
Rocky and Bullwinkle	84	quartz
Roy Rogers	85	Bradley/with Dale Evans/Happy Trails
Sam the Eagle	84	Olympic/metal case
	84	Olympic/plastic/fliptop
Scooby Doo	78	Picco/brown plastic
Skippy	80s	moving eyes
Sluggo	74	Bradley
Smokey Bear	79	Bradley/rotating bezel & date
Smurf	83	Bradley/animated hands/blue and yellow
Smurfette	83	Bradley/animated hands/blue and yellow
Snoopy	70s	run-away/blue face
	76	blue backgrnd/white Snoopy
	76	blue denim backgrnd/playing tennis
	77	playing tennis/yellow
	77	red backgrnd/Woodstock second-hand
	79	Snoopy on house/black/Flying Ace
Speedy Gonzales	70s	Hong Kong
Spiderman	78	Dabs
Star Wars	77	Darth Vader/metal case
	77	R2D2 3CPO/blue sky
	77	R2D2 brown
	77	R2D2 German
	79	Darth Vader/black and silver band
	79	R2D2 3CPO/black and silver band
	79	Wicket/black and silver band
	79	Yoda/black and silver band
	83	Jabba/small/silver
	83	Jedi/small/silver
	83	Yoda/small/silver
Strawberry Shortcake	80	Bradley/round
Superman	78	Dabs
	78	Timex
Sylvester	85	blue plastic/Hong Kong
Tarzan	83	Bradley/ape
	83	Bradley/tree
Tigger	80s	small size
Tom and Jerry	85	Bradley/Jerry on water spout
	85	Bradley/white plastic/in heart
Tom Thumb	80s	World-wide Watch Company
Touche Turtle	80s	Hong Kong
Ugly Duckling	80s	yellow face
Underdog	84	A & M
Winnie the Pooh	77	digital
	77	Sears/Honeypot
	80s	Bradley/blue
Wonder Woman	78	Dabs
	78	Timex
Woodsey Owl	70s	Smokey's friend
Woody Woodpecker	78	Bradley/with Sylvester
	80s	Hong Kong
Wuzzies	85	Bradley

Chapter Six
The Excitement Continues:
1985-The Present

In 1986 the Seiko Watch Company was awarded the license for Disney watches. Seiko continued with Bradley's marketing theory--produce many different models of each character every year. While Seiko and Lorus, their subsidiary, are comparatively new players in the comic character watch market, there are already some models I consider to be extremely collectible. The 1986 metal Mickey and Donald, the first watch to picture both characters, sold on the secondary market for five times its retail cost. Other collectible pieces include the 60th anniversary watches, the Mickey Hollywood series, and all models related to the Fantasia movie.

During this time, various companies were given the rights to produce watches exclusively for the Disney parks or for Disney employees. Many of these rare watches such as the Mickey policeman, the Mickey fireman, or Mickey as a maintenance worker would be a find in anyone's collection. There were two interesting art deco watches available only at the parks. One has Mickey in silhouette, and the other has Mickey looking at the movement which is visible from the front.

There were new developments during this period such as the hologram watch. These included Mickey astronaut, Mickey's face, Mickey Fantasia, Bugs Bunny 50th Anniversary, and Fred Flintstone. The company Upper Deck, better known for its baseball cards, issued a series of Looney Tune hologram watches featuring characters in various major league baseball uniforms.

Great new characters were introduced during the mid-1980s. The first watch issued of any new character has the potential of becoming extremely collectible. Examples include Roger Rabbit (three different models), Jessica Rabbit, and the limited edition Teenage Mutant Ninja Turtles. Armitron has issued a series of watches over a three-year period which I believe every collector should acquire. The series includes Bugs Bunny, Roger Rabbit, Jessica Rabbit, the Roadrunner, Porky Pig, Elmer Fudd, Garfield, Snoopy, and others.

The 1980s and 1990s appear to be the era of limited edition watches. Teenage Mutant Ninja Turtles (mentioned above), Paddington Bear, Uncle Scrooge, Mickey Fantasia, and the reissuing of Mickey #1 are only a few of these. The most famous of the limited editions to date, however, is the gold Pedre backwards Goofy, a copy of the 1971 Goofy. In September 1990, the Disney catalog listed this watch for $75 and sold out within two days. Within three months, it was available at watch shows for $500! Because of its success, Pedre introduced a $75 silver version in a non-limited edition issue which also sold well. In the following year, Lorus made both a gold and silver small-size version for $29.95. At the present time, the limited edition 5000 gold Pedre backwards Goofy is still selling for $300.

Archie, 1991. Size M.

Archie, 1990. Size L.

Barbie, 1990, Armitron. Size M.

Beany and Cecil, 1990. Made for the fan club. Size M.

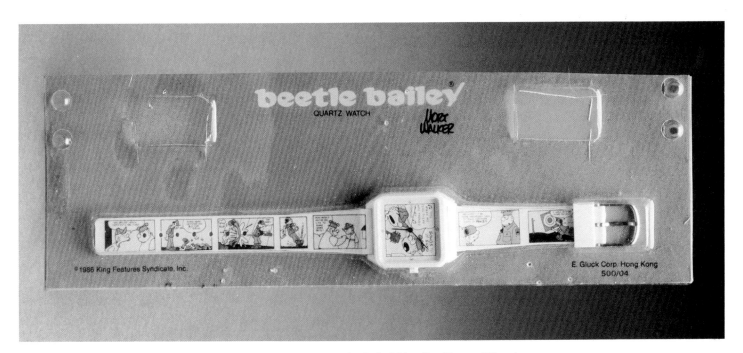

Beetle Bailey, 1986 series that included Blondie, Nancy, Hi and Lois, and Hagar the Horrible. Size M.

This latest period has seen the introduction of plastic molded watches which are displayed in blister-packs with delightful graphics. Lift up the figure and see the LCD tell time. You can begin the collection with the 21 Teenage Mutant Ninja Turtles, four of which are talking watches. Then add the Duck Tales, Rescuers Down Under, Tale Spin, Jungle Book, Tiny Toon, Darkwing Duck, and, my favorites, the Marvel Comic's Captain America, Punisher, Spiderman, and the Hulk. A note of caution to the collector who wants to acquire them all: on the reverse side of the packaging is normally a picture of all the watches to be included in that series. I have spent many hours searching for a watch that was never issued.

In early 1991 Disney began issuing watches for sale only through their catalog or through the Disney Store. These included the Snow White 50th Anniversary and the reissue of the Mickey #1 in addition to others. If not in limited editions, these watches are normally made in limited quantities. I suggest making friends with employees of the nearest Disney Store to keep informed of the latest watches available. In addition Disney now makes limited edition watches for sale only to Disney employees. Watches such as the Goofeteer Tinkerbell, Rocketeer and McTracey have become extremely expensive on the secondary market, selling for as much as $300 contrasted with the employee cost of somewhere between $30 or $70.

Comic Time is a new company who is producing primarily limited edition watches which are sold only through their club concept. Upon joining the club, a subscriber receives a watch every six weeks. The watches are packaged in tin toy containers which are collectible themselves. The watches are a throw-back to the old days featuring mechanical, wind-up movements, in lieu of the quartz to which we have become so accustomed. Comic Time's first four clubs include Marvel characters such as Spiderman and Captain America; DC Comics characters such as Superman and the Flash; all Star Trek characters including the Next Generation; and finally, the series of King Features characters of both old and new comic strip personalities such as Flash Gordon, Barney Google, Prince Valiant, Katzenjammer Kids, Dennis the Menace, Beetle Bailey, Zippy, and Hi and Lois.

By the time you read this book, there will have been many more watches worthy of having been included in this chapter. More importantly, however, I believe that there is a resurgence of producing and appreciating high quality collectible comic character watches. To the collector, the future is exciting.

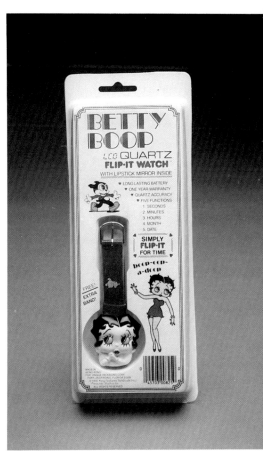

Betty Boop, one of a 1986 series that included Tweety, Popeye, Sylvester, Daffy Duck, and Bugs Bunny. Size M.

Big Bad Wolf, 1991, Lorus revolving disc watch. One of a series of two. Size M.

Big Bad Wolf, 1991, Lorus revolving disc watch. Part of a series of two. Size M.

Blondie, one of the 1986 series that include Beetle Bailey, Hagar, Nancy, and Hi and Lois. Size M.

Brother Bear, 1986. An early version of an Armitron character watch. Size M.

Bugs Bunny, one of the 1986 series that included Tweety, Popeye, Betty Boop, Sylvester, and Daffy Duck. Size M.

Bugs Bunny, 1990, Armitron. Part of a series of sixteen. Size M.

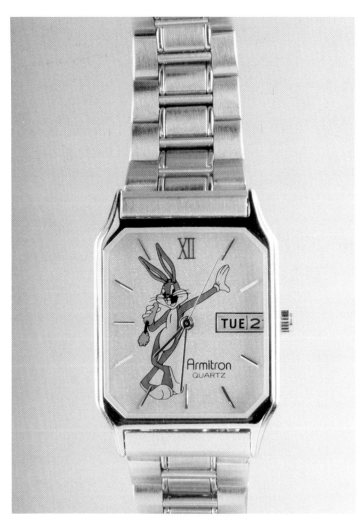

Bugs Bunny, 1991, Armitron. Size M.

Bugs Bunny, 1990, Armitron. 50th Anniversary watch with diamond chip at number 6. Size M.

Bugs Bunny, 1991, Armitron. Part of a series of sixteen. Size M.

Bugs Bunny, Armitron, 1991. 50th Anniversary edition. Part of a series of sixteen. Size M.

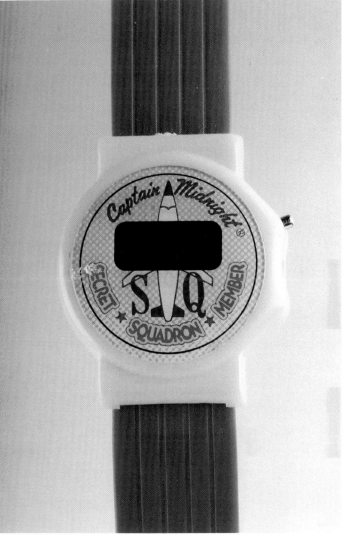

Captain Midnight, 1988 model made as a promotional for Ovaltine. This is the only Captain Midnight watch made. Size M.

Bugs Bunny. Part of a series of what was supposed to be 26 baseball watches with a Looney Toon character representing each team. These were hologram watches made by Upperdeck, however the series was discontinued after only fourteen watches were made. Size M.

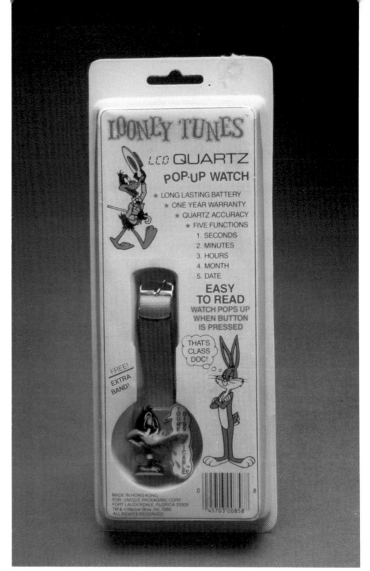

Daffy Duck. part of a 1986 series that include Tweety, Popeye, Betty Boop, Sylvester, and Bugs Bunny. Size M.

Curious George, 1986, Hong Kong version. Size S.

Daffy Duck, 1990, Armitron. Part of a series of sixteen to date. Size M.

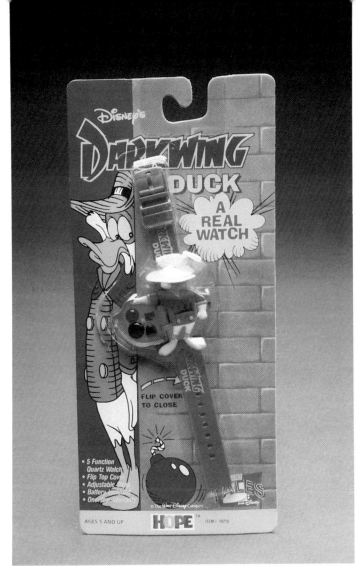

Darkwing Duck, 1991, by Hope. Size M.

Ducktales, 1988, Lorus. Size M.

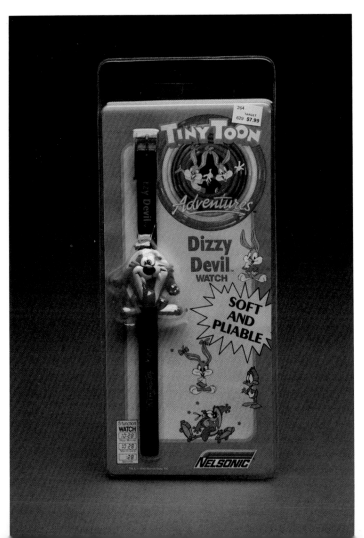

Dizzy Devil. Part of a series of four of Tiny Toon characters. Size M.

January 1991 New Model Information

LORUS®

LFDI27R-E (Red)
3.9" H × 7.6" W × 2.6" D
Disney melody alarm clock featuring four separate melodies: It's A Small World, Mickey Mouse March, When You Wish Upon A Star, Electric Parade. Pop up alarm with second hand. 3 "AA" batteries (not included).

NEW

LFDI27W-E (White)
3.9" H × 7.6" W × 2.6" D
Disney melody alarm clock featuring four separate melodies: It's A Small World, Mickey Mouse March, When You Wish Upon A Star, Electric Parade. Pop up alarm with second hand. 3 "AA" batteries (not included).

NEW

LFDI28R-E
5.0" H × 6.8" W × 2.4" D
Action circus alarm clock featuring Mickey, Minnie and Goofy as circus performers. Action dial with constant motion as Goofy juggles. Red case with white dial. Pop up alarm with second hand. "AA" and "C" batteries (not included).

NEW

LFDI25H-A
3.0" H × 3.5" W × 1.6" D
Mickey Mouse alarm clock featuring a close-up view of Mickey's face. Pop up alarm and second hand. Black case. "AA" battery (not included).

NEW

LFW223F-A
10.4" Round × 1.5" D
"Duck Tales" wall clock featuring the popular stars of the children's TV show. Second hand. Red case with white dial. "AA" battery (not included).

NEW

LFW227R-E
13.2" H × 8.6" W × 1.8" D
Mickey Mouse pendulum wall clock. Pendulum provides constant action as Mickey "kicks" the soccer ball. Red case, clear plastic face and pendulum. 2 "AA" batteries (not included).

NEW

LORUS
Clock Division
Seiko Corporation of America
1111 Macarthur Boulevard
Mahwah, New Jersey 07430

© The Walt Disney Company

Ref. No. CCS-91
Printed in U.S.A.

Disney Watches from HOPE!

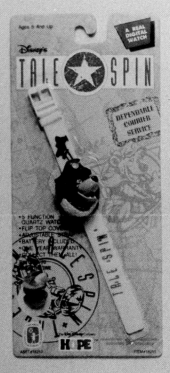

TaleSpin
Molded Watch
Assortment #18210
Pack: 24

**The Rescuers
Down Under**
Molded Watch
Assortment #17210
Pack: 24

**Chip 'n Dale
Rescue Rangers**

Molded Watch
Assortment #15210
Pack: 24

DuckTales
Molded Watch
Assortment #14210
Pack: 24

These watches also available in **Disney Assortment #10210**. Pack: 24

Elmer Fudd, 1991, Armitron. Part of a series of sixteen. Size M.

Garfield, made for Armitron, 1990. Part of series of sixteen. Size M.

Fred Flintstone, 1991, extra large size.

Garfield, 1990, Armitron. This gold relief dial was placed on three watches. Snoopy and Kermit were the other two. Size M.

Garfield, 1991, Armitron. Part of a series of sixteen. Size M.

Goofy, 1991, Lorus and Pedre silver backwards Goofys. Size M.

Goofy, made by Pedre, 1990. This is a Limited Edition of 5000 and is a reissue of the 1971 Helbros Goofy. This watch sold for $75, and all 5000 were purchased within nine days of issue. Within 6 months it was selling for $500 at collectible shows when Pedre reissued it in silver in a non-limited edition for the same $75 price. The following year Lorus issued a smaller version in gold and silver for sale for $29.95.

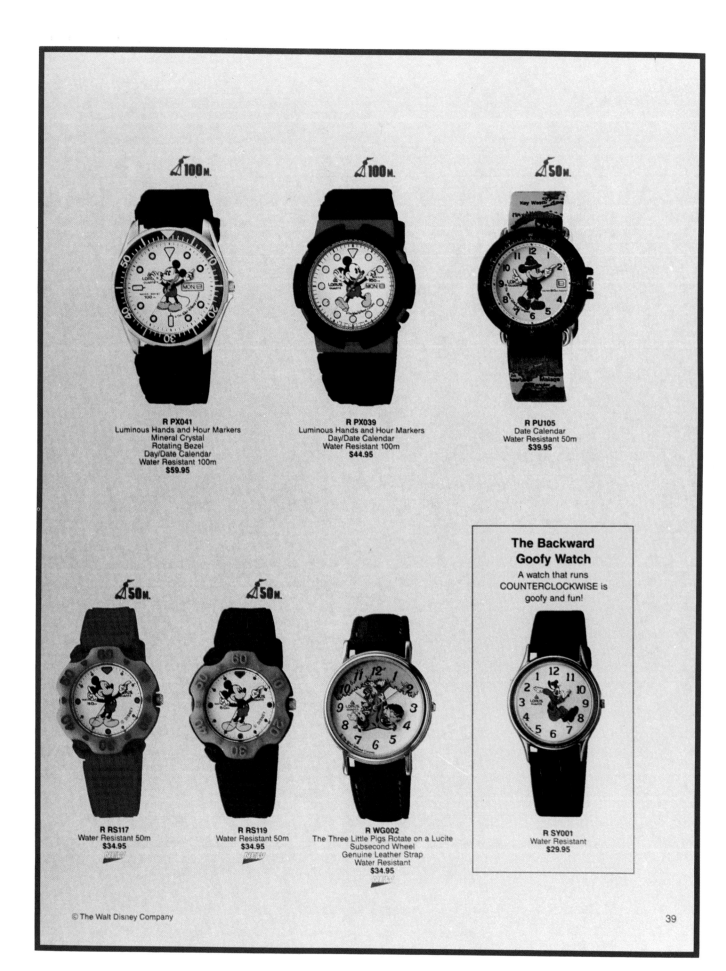

R PX041
Luminous Hands and Hour Markers
Mineral Crystal
Rotating Bezel
Day/Date Calendar
Water Resistant 100m
$59.95

R PX039
Luminous Hands and Hour Markers
Day/Date Calendar
Water Resistant 100m
$44.95

R PU105
Date Calendar
Water Resistant 50m
$39.95

R RS117
Water Resistant 50m
$34.95

R RS119
Water Resistant 50m
$34.95

R WG002
The Three Little Pigs Rotate on a Lucite
Subsecond Wheel
Genuine Leather Strap
Water Resistant
$34.95

The Backward Goofy Watch

A watch that runs COUNTERCLOCKWISE is goofy and fun!

R SY001
Water Resistant
$29.95

Gumby and Pokey, 1986, quartz. Size M.

Howdy Doody, 1987, Concepts Plus. 40th Anniversary edition. Size M.

Howdy Doody, 40th Anniversary clock. *Courtesy of Jack Feldman.*

Jessica Rabbit, 1987, by Amblin. Size M.

Jessica Rabbit, 1991, Armitron. Part of a series of sixteen. Size M.

Jessica Rabbit, 1991, for sale at the Jessica Store, Walt Disney World, Orlando, Florida. Size M.

Jessica Rabbit. Made for sale at the Jessica Store only at Pleasure Island at Disney World in Orlando, Florida. Jessica's leg is animated and kicks back and forth. Size M.

Jetson's, 1991, with revolving disc. Size M.

Kermit, 1990, Armitron. Part of a series of sixteen to date. Size M.

Kermit, 1987. Given to show production members. Size M.

Little Mermaid, Part of the 1990 series including Ariel, Sebastian, and Flounder. Size M.

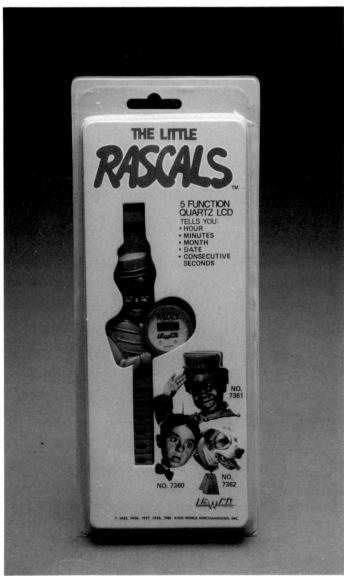

Little Rascals, 1987, plastic model. Size M.

Looney Toons, 1991, Armitron. This is the first of Armitron's new concept of 3-D watches. The hands of the watch are behind Bugs, but are in front of Sylvester and Daffy. Others in the series are Popeye, Tom and Jerry, Dagwood and Blondie. Size M.

Looney Toons, 1990, Armitron. One of a series of sixteen. Size M.

Mad, 1987, made for Concepts Plus. Size M.

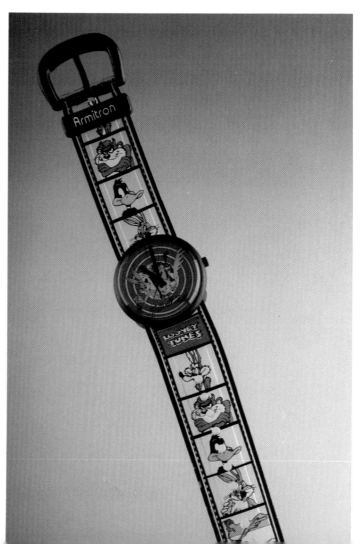

Looney Toons. An interesting new concept in band design by Armitron. Size M.

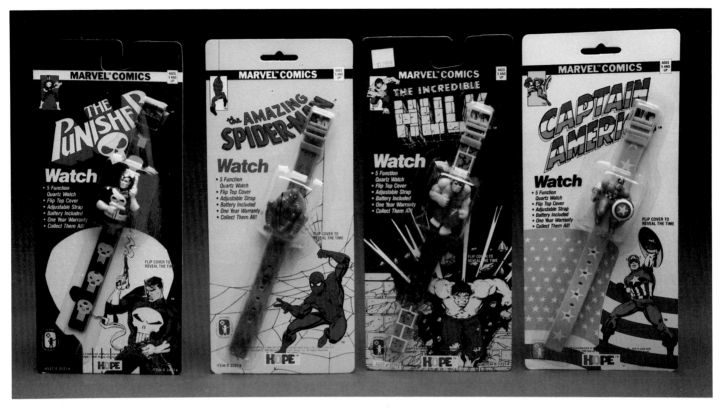

Marvel Comics. Various plastic molded Marvel comic characters by Hope Industries, 1990. Size M.

Mickey and Donald. Made by Lorus for only one year and then discontinued. When collectors realized that this was the only Mickey and Donald watch, the value sky-rocketed. Size M.

Mickey and Minnie, 1990, Lorus. Mickey and Minnie in love. Size M.

Mickey Mouse, Goofy, and **Donald Duck**, 1990, Lorus. The crystal on these watches is flat. In future years they changed to a rounded crystal. Size M.

Mickey Mouse, 60th Anniversary, 1988, by Seiko. This commemorates the first Mickey Mouse cartoon. Size M.

Mickey Mouse. 60th Anniversary prototype watch which was never mass produced. Size M.

The Seiko Disney Collection

QFD205B

QFD206G

QFD205B
Voice and music alarm.
8½" x 7¼" x 3¼"

QFD206G
Sing-Along Alarm
Choice of 7 Disney songs or a beep:
— The Mickey Mouse Club March
— It's a Small World
— Supercalifragilisticexpialidocious
— Heigh-Ho
— Zip-a-dee-doo-dah
— When You Wish Upon a Star
— Bibbidi-Bobbidi Boo
  (The Magic Song)
— 7¼" x 5⁷⁄₁₆" x 3¹⁄₁₆

© The Walt Disney Company

Sing Along Alarm

QQM122B
Choice of 6 songs or a beep:
— Piano Man
— New York, New York
— Stand By Me
— Here Comes The Sun
— Get Happy
— The Entertainer
7¼" x 5⁷⁄₁₆" x 3¹⁄₁₆"

QQM122B

Mickey Mouse. The 60th Anniversary watch by the Japanese company Alba to celebrate the 1928 Mickey Mouse movie "Steamboat Willie." Size M.

Mickey Mouse. The Lorus version of the 60th Anniversary of Mickey in 1988. Size M.

The Seiko Disney Collection

QFW101W

SEIKO MAGICAL, MUSICAL CLOCK

Every hour, on the hour* this unique wall clock plays one of six Disney tunes. As it plays, each panel swivels open to reveal a Disney character. The cast of characters includes: Mickey, Minnie, Goofy, Daisy Duck, Donald Duck and Pluto. Its repertoire of six songs consists of: Heigh Ho, The Mickey Mouse Club March, Supercalifragilisticexpialidocious, Zip-A-Dee-Doo-Dah, When You Wish Upon a Star and Bibbidi Bobbidi Boo. When the song is over, the clock strikes the hour and all panels close to the tune of "It's a Small World."

*Clock has a built-in nighttime silencer. A light sensor disengages all music, chimes and movement.

QFD210G
Hand-painted ceramic.
6" x 7¼" x 2½"

QFD210G

6

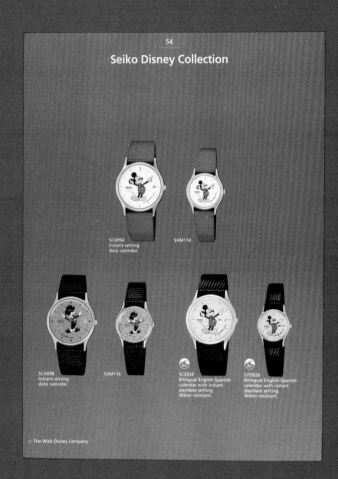

54

Seiko Disney Collection

SCX094
Instant-setting
date calendar.

SXM114

SCX098
Instant-setting
date calendar.

SXM116

SCZ034
Bilingual English-Spanish
calendar with instant
day/date setting.
Water-resistant.

SYD026
Bilingual English-Spanish
calendar with instant
day/date setting.
Water-resistant.

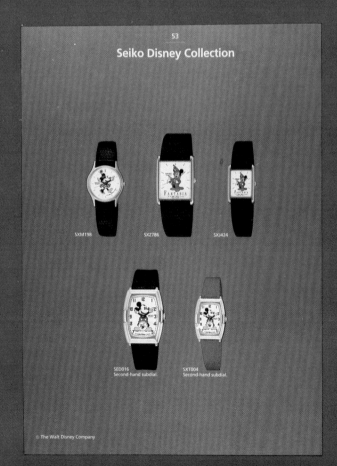

53

Seiko Disney Collection

SXM198

SXZ786

SXJ424

SED016
Second-hand subdial.

SXT004
Second-hand subdial.

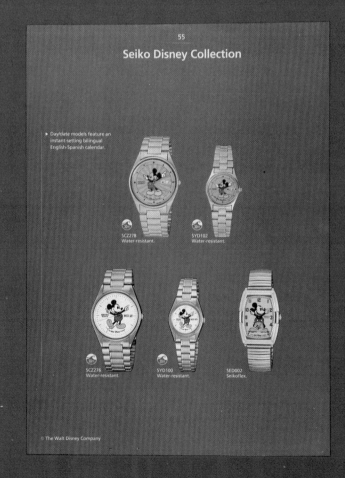

55

Seiko Disney Collection

▶ Day/date models feature an
instant-setting bilingual
English-Spanish calendar.

SCZ278
Water-resistant.

SYD102
Water-resistant.

SCZ276
Water-resistant.

SYD100
Water-resistant.

SED002
Seikoflex.

Mickey Mouse,1989, Pedre. Mickey Astaire is a companion piece to Minnie Marilyn. Size M.

Mickey Mouse. The 1990 version of Mickey Hollywood. One of seven in which Mickey is wearing long pants. Size M.

Mickey Mouse, 1990 Lorus version of Mickey Fantasia. Size M.

Mickey Mouse, 1990. A version of Mickey Fantasia by Seiko. Size M.

Mickey Mouse, 1989, by Breil. Part of a series of two, the second of which is Mickey next to an airplane. Size M.

Mickey Mouse, 1990, Fantasma. One of a series of six hologram watches made by this company. Size M.

Mickey Mouse, 1990, Lorus melody watch. Size M.

Mickey Mouse, 1990, Lorus. Sometimes called the Mickey Rolex. Size M.

Mickey Mouse, 1990, Lorus. Known as Mickey Hollywood. Size M.

Mickey Mouse, 1990, made for the Disney Catalog. Size M.

Mickey Mouse, 1990. A Lorus copy of the 1933 Mickey. Size M.

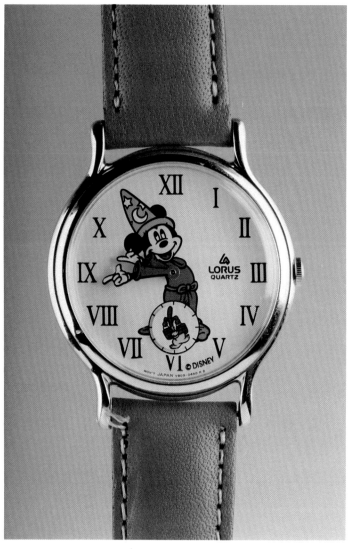

Mickey Mouse, 1991, Lorus version of Mickey Fantasia. Size M.

Mickey Mouse. 1990 Walt Disney World Teddy Bear Convention. This watch was sold only at the convention and its companion piece was Minnie Mouse and a doll. Size M.

Mickey Mouse. Mickey Mouse Grad Night watches were for sale only at Disneyland in California and were discontinued in 1990. Size M.

Mickey Mouse, 1991, Lorus. Size M.

Mickey Mouse. Made for Disney employees only in 1991, it is referred to as the McTracy watch. Size M.

Mickey Mouse. Pulsar's first Mickey Mouse watch. It came in three sizes: small, medium, and large.

Mickey Mouse. Self-portrait watch made in 1991. It will obviously become collectible because it is the only watch that pictures not only Mickey, but also Walt Disney's face. Size L.

Mickey Mouse. 1991 Silhouette for sale at Disney Parks. Size M.

Mickey Mouse, 1991, Lorus. Called "Mickey Through the Years," as the disc rotates we see different versions of Mickey. Size M.

Mickey Mouse, 1991, Fantasia 50th Anniversary limited edition of Mickey shaking Stokowsky's hand. This watch did not sell well because people did not recognize director of the music for Fantasia. Size M.

Mickey Mouse, 1991, Lorus. Mickey Hollywood. One of seven with Mickey in long pants. Size M.

Mickey Mouse. One in the series made for sale to employees at the Disneyland parks. The others include Mickey as a fireman and as a sanitation warker. Size M.

Mickey Mouse. Another of the watches made for sale to employees who are part of the firefighting team at Disney parks. There are companion pieces with Mickey as a policeman and as a sanition worker. Size M.

Mickey Mouse, Limited Edition Fantasia watch. Part of a series that included Dumbo and Uncle Scrooge. Size M.

Minnie Mouse. Minnie Marilyn is a companion piece to Mickey Astaire made in 1989 by Pedre. Size M.

Mighty Mouse, 1985, Bradley. Size M.

Minnie Mouse, 1990. Disney World Doll Convention. This watch was only sold during the convention and the companion piece was Mickey and a teddy bear. Size M.

Monopoly, 1986, Armitron plastic. Size M.

Muffy Vanderbear, 1990. The official mascot of bear collectors. Size M.

Mouse Club. Symbol of the Mouse Club Collectors of Disneyana. The Disney company would not allow these collectors to use a picture of Mickey Mouse, so they put this bag over Mickey's head and use it as their symbol. Size M.

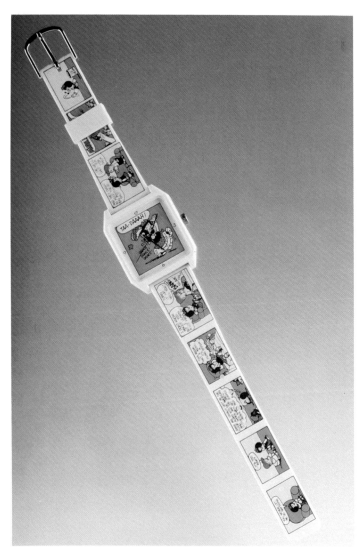

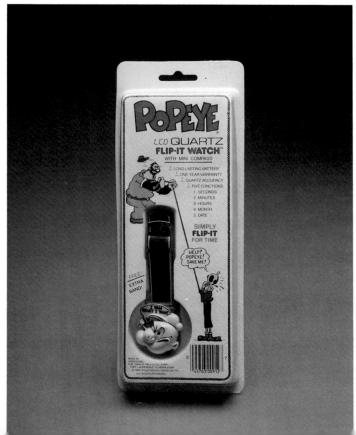

Nancy, made by Sutton Time. This is part of series of cartoon strip watches which included Hagar the Horrible, Hi and Lois, Beetle Bailey, and Blondie and Dagwood. Size S.

Pee Wee Herman, 1989, by Nelsonic. Size M.

Pink Panther, 1991, Armitron. Part of a series of sixteen watches. Size M.

Popeye, one of a 1986 series of six watches that included Tweety, Betty Boop, Sylvester, Daffy Duck, and Bugs Bunny. Size M.

Porky Pig, 1991, Armitron. Part of a series of sixteen. Size M.

Roger Rabbit, 1987, by Amblin. One of a series of three. Size M.

Roadrunner, 1991, Armitron. Part of a series of sixteen. Size M.

Roger Rabbit, 1987, Hong Kong version. Size M.

Roger Rabbit, 1987, by Amblin. One of series of three. Size M.

Roger Rabbit, 1990, Armitron. Part of a series of sixteen to date. Size M.

Roger Rabbit, 1987, by Amblin. One of a series of three. This one is called "The Silhouette." Size M.

Snoopy, 1991, Armitron. Part of a series of sixteen. Size M.

Snoopy, 1991, Joe Cool. Size M.

Snoopy, 1990, Armitron. This gold relief dial is part of a series that included Garfield and Kermit. Size M.

Snow White. Made for Disney, 1991, Limited Edition. This watch has a revolving disc which makes it appear that Snow White is kissing each character. Size M.

Splash Mountain, 1989. A waterwatch sold at Disneyland to commemorate the Splash Mountain ride. Size M.

Superman, 1986. The Smithsonian issued this 50th Anniversary Superman watch. Size M.

Spock, Jetsons, and **Flintstone,** 1990. Spock is from a series of five Star Trek watches produced by Bradley. Size M.

Sylvester, from a 1986 series that included Tweety, Popeye, Betty Boop, Daffy Duck, and Bugs Bunny. Size M.

Sylvester. The 1986 version of the Hong Kong plastic watch. Size M.

Teenage Mutant Ninja Turtles, 1991, Hope Industries. Part of a series of four talking watches. Size M.

Watches • Sneaker Snappers™ • Calculators • Strapper Snapper™ • Keychains • Gift Sets
Magnets • Wall Watch™ • Alarm Clock • Shoe Laces • Cling-It™ • Hook-It™ • Wrist Snapper™

Heroes in a half shell™

NEW!

MIKE.™
THE SEWER
SURFER

LEO,™
THE SEWER
SAMURAI

RAPH,™
THE SPACE
CADET

THE
UNDERCOVER
TURTLE

NEW!

SPLINTER™   APRIL O'NEIL™   SHREDDER™

#60214
Children's
Molded Watches
With Flip Top Cover
Assortment
Pack: 24
Blister Carded

Watches

"One good turtle deserves another."

#60213 **Interchangeable Lens Watch**
Pack: 24

#60510 **Water Resistant Sports Watch**
Pack: 24 Blister Carded

#60614 **Boy's Analog Plastic Watch**
Green or Black Band
Silver or Gold Face
Pack: 6

#60714 **Men's Analog Dress Watch**
Gift Box Pack: 6

#60214Q **Limited Edition Gold Plated Quartz Watch**
Gift Box Pack: 1

Alarm Clock

#60790 **Desk/Alarm Clock**
Blister Carded
Pack: 12

Teenage Mutant Ninja Turtles, 1990. Limited Edition by Hope
Industries, a companion piece to Paddington Bear. Size M.

Teenage Mutant Ninja Turtles, 1990, Hope. Part of seventeen
watches in the series. Size M.

Tweety, 1991, Armitron. Part of a series of sixteen. Size M.

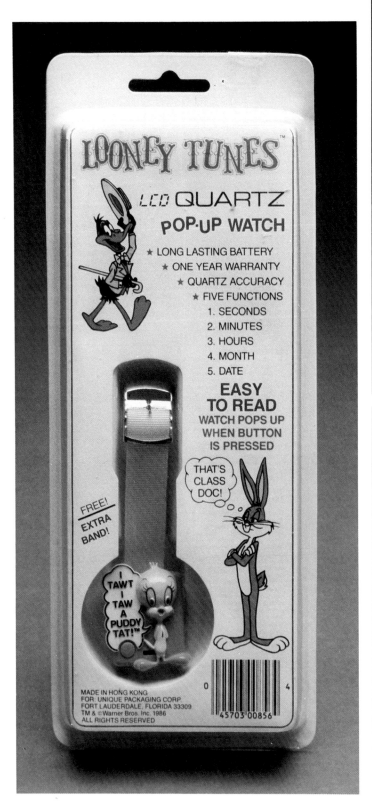

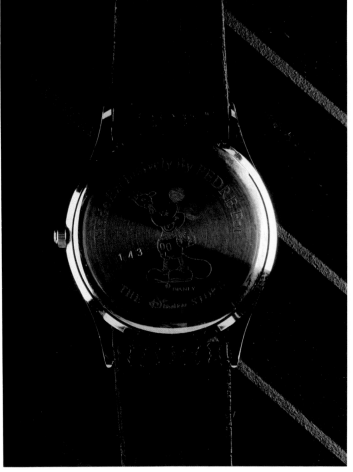

Tweety, part of a 1986 series of six that included Sylvester, Popeye, Betty Boop, Daffy Duck, and Bugs Bunny. Size M.

Uncle Scrooge. 1991 Disney Limited Edition watch. This is part of a series that included Fantasia and Dumbo. Size M.

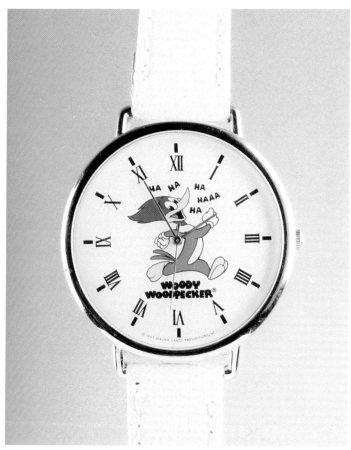

Walt Disney World. One of only two models that have Mickey and Donald on the same dial. Size M.

Woody Woodpecker, 1991, extra large size.

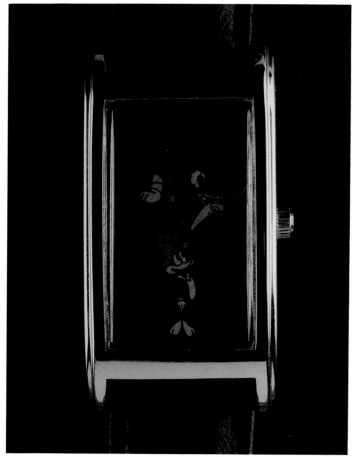

Winnie the Pooh, 1991, Lorus. This watch has a revolving disc. Size M.

Woody Woodpecker. 1991 Limited Edition to celebrate Woody's 50th anniversary. Size M.

Woody Woodpecker, 1991, 50th Anniversary of Woody
Woodpecker. Size M.

Yosemite Sam, 1991, Armitron. Part of a series of sixteen. Size M.

Woody Woodpecker, 1991 versions of Woody. Size M.

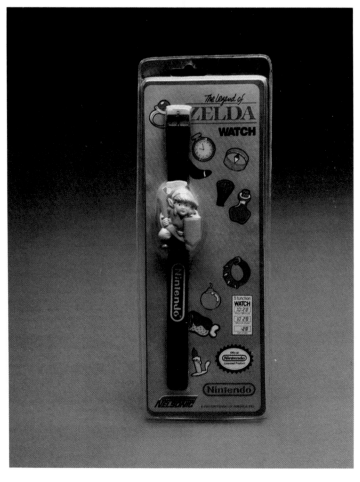

Zelda, 1990, Nelsonic. Companion to Super Mario. Size M.

THE EXCITEMENT CONTINUES: 1985-THE PRESENT

WRISTWATCHES

NAME	YR	NOTES
Alf	80s	furry
Alfalfa	86	plastic molded/LCD
Alvin	90	Armitron/red plastic
American Tail	91	Tiger/prototype
	91	Wyle Burp/prototype
An Pan Man	91	Japan/molded plastic
Archie	80s	revolving disc/Betty & Veronica
	89	airplane
Astro Boy	90	color photo
	90	with friends
Baby Face	91	So Happy Heidi
	91	So Loving Laura
	91	So Sorry Susan
	91	So Surprised Suzie
Bad Boy	90	Sutton Time/plastic case/graphic band
Barbie	80s	Armitron
Batman	80s	Japanese/plastic molded/LCD
	88	card
	89	flying
Beach Network--Bad Spike	91	plastic molded/fliptop
Beach Network--Beach Creach	91	plastic molded/fliptop
Beach Network--Skate Rat	91	plastic molded/fliptop
Beanie and Cecil	90	Collector's Fan Club
Beauty and the Beast	91	plastic/pop-up
Beetle Bailey	86	cartoon strip
Beetlejuice	89	plastic/fliptop/LCD
Betty Boop	91	moving eyes
Big Bad Wolf	91	Lorus/wolf on revolving disc
	91	wolf behind tree/pigs on revolving disc
Blondie	86	cartoon strip
Bozo the Clown	80s	plastic molded/fliptop/LCD
Breathless Mahoney	90	Canadian/changeable faces
	90	Timex
Buckwheat	86	bendable plastic/LCD
	86	plastic/fliptop/LCD
Bugs Bunny	80s	Armitron/gray body/carrot in hand
	86	plastic/fliptop/LCD
	90s	Armitron/black tux
	90s	Armitron/blue tux
	90	50th hologram anniversary
	90	Armitron/50th anniversary/gold with diamond chip
	90	flipover/baseball/LCD
	90	Sutton Time/plastic case/graphic band
	91	Armitron/gold/metal/day & date/rectangular
Captain America	90	plastic molded body/fliptop/LCD
Captain Midnight	88	Led/made for Ovaltine
Captain Planet	91	plastic molded/LCD
Chibi	90	Japanese/girl/cartoon
Chip and Dale	90	small/black plastic
	91	plastic molded/fliptop/LCD
Chipmunks	90	red/digital
Daffy Duck	86	plastic/fliptop/LCD
	91	Armitron
Dalmatians--Patch	91	no dots/yellow ear/plastic molded
Dalmatians--Perdita	91	dots/pink ear/plastic molded
Dalmatians--Pongo	91	black eye/yellow ear/plastic molded
Darkwing Duck	91	plastic molded/LCD/prototype
Dick Tracy	90	Canadian/changeable faces
	90	England/black/sport
	90	English/talking
	90	metal/radio watch/in movie
	90	movie role watch/not issued
	90	plastic/speaker/digital
	90	radio/fliptop/French-Canada
	90	talking watch
	90	Timex/black and blue profile
	90	Timex/white profile
Disneyland	90	35th anniversary/molded band
Donald Duck	80s	Lorus/large size/flat crystal
Doonesbury	91	Duke watch
Dr. Man	80s	Japanese/plastic molded/LCD
Dr. Spock	86	bendable/plastic/LCD
Duck Tales	88	Huey/Duey/Luey
Duck Tales--Duey	91	plastic molded/fliptop
Duck Tales--Huey	91	plastic molded/fliptop
Duck Tales--Launchpad	91	plastic molded/fliptop
Duck Tales--Luey	91	plastic molded/fliptop
Dumbo	91	Alba/pink plastic
E.T.	90	blue plastic
Elmer Fudd	90s	Armitron
Felix the Cat	90	black and white/moving eyes
	90	black metal box
Fido	90	black and white silhouette
Fievil Mouse	90	Universal Studios
Flattop	90	Timex/"Eat Lead"
Flintstones	91	Bedrock/moving dial
Fred and Dino	86	bendable/plastic/LCD
Fred Flintstone	90	large size
	91	hologram
	91	plastic/molded/LCD
Garfield	80s	Armitron/standing
	80s	plastic molded/LCD
	80's/furry	
	90s	Armitron/blue backgrnd/jumping over moon
	90	Armitron/gold embossed
	90	plastic/tell time
	91	Armitron/LCD
George and Astro	86	bendable/plastic/LCD
Ghostbusters	89	plastic molded
	89	Slime/plastic molded
	89	Stapuff Marshmellow Man/plastic molded
Gobot	80s	red/digital
Godzilla	80s	Japanese/plastic molded/LCD
Goofy	80s	Lorus/large size/flat crystal
	90	Lorus/silver/backwards
	90	Pedre/gold/backwards/limited edition
	90	Pedre/silver/backwards
	91	Lorus/gold/backwards
Gremlin	90	Gizmo/blue plastic molded/LCD
Gumby	86	bendable plastic/LCD
Gumby and Pokey	86	quartz
Gundam	80s	Japanese/plastic molded/LCD
Gyavan	80s	Japanese/plastic molded/LCD
Hachman	80s	Japanese/plastic molded/LCD
Howdy Doody	87	40th anniversary/metal/green face
	88	40th anniversary/metal/Howdy at 3-6-9-12
	88	40th anniversary/red plastice/green face
Hulk	91	plastic molded/LCD
Jessica Rabbit	87	Amblin/5-colors
	87	Armitron
	91	Amblin/moving foot
	91	Red face
Jetsons	90s	Elroy and Astro/plastic molded/LCD
	90s	Judy/plastic molded/LCD
	90	white revolving face
Joker	80s	DC Comic/green face
	88	card
	89	white face/diamond
Judy Jetson	86	bendable/plastic/LCD
Jughead	80s	revolving disc
Jungle Book Characters	91	plastic molded/fliptop
Jungle Book--Baloo	91	plastic molded/fliptop
Jungle Book--Mawgli	91	plastic molded/fliptop
Jungle Book--Shire Khan	91	plastic molded/fliptop
Kamen Rider	80s	Japanese/plastic molded/LCD
Kermit the Frog	80s	Armitron
	87	face only
King Kong	90	black plastic
Kit Kat	89	black and white
Little Mermaid--Ariel	90	mermaid floating in water
	90	molded plastic
	90	pink band
Little Mermaid--Flounder	90	molded plastic
	90	yellow band
Little Mermaid--Sebastian	90	crab/red band
	90	molded plastic
Little Pony	91	red/plastic molded
Looney Tunes	90s	Armitron/6 characters
	90	Sutton Time/filmstrip
	90	Sutton Time/plastic case/graphic band/all characters
	91	Armitron
	91	limited edition
MAD	87	Alfred E. Neuman/limited edition
Mickey and Bear	90	Walt Disney World Bear Convention
Mickey and Donald	78	Lorus/metal band
	90	for Walt Disney World
Mickey and Minnie	91	On the Moon/Fossil
Mickey and Pluto	90	white
Mickey Mouse	80s	Disney Channel
	80s	Disney Channel promo
	80s	Lorus/Fantasia/black plastic
	80s	Lorus/large size/flat crystal
	80s	policeman
	87	Mickey white silhouette
	88	60th anniversary/Alba/Steamboat Willie
	88	60th anniversary/colored Lorus version
	88	Breil/Mickey & Minnie at car
	88	Lorus/60 Years/gold writing on silver face
	88	Lorus/copy of Mickey One
	88	Seiko/Mickey Rolex
	88	Seiko/repro of 1938 60th birthday
	89	Hollywood/in tux
	89	Lorus/alarm/melody
	89	Seiko/Mickey tux
	90	50th anniversary Fantasia/black face
	90	Briel/by airplane
	90	by station wagon/plastic
	90	Fantasia/gold/limited edition
	90	Fantasia/hologram
	90	Fantasia/shaking hands with Stokowski
	90	Fantasia/white face
	90	golf club/red plastic
	90	golfer in color
	90	Hollywood filmstrip
	90	Lorus/Fantasia/revolving disc
	90	Lorus/hands open
	90	Lorus/large
	90	melodies/silver
	90	Pedre/Astaire
	90	Pedre/copy of 1933 with metal band
	90	Pulsar/black silhouette
	90	special edition/firemen
	90	special edition/graduate
	90	team sport/yellow
	91	2-gun cowboy
	91	2-songs melody
	91	Christmas/Nutcracker/limited edition
	91	circular disc/Mickey & Pluto/Fossil
	91	Donald, Mickey, Goofy/Fossil
	91	face hologram
	91	faces of Mickey/Fossil
	91	Fantasia/hologram
	91	forest ranger
	91	Fossil/skeleton
	91	friends/small size
	91	gold-embossed/golfer
	91	McTracy
	91	Pedre/limited edition/leather band
	91	Pulsar/large black
	91	Pulsar/large/white face
	91	self-portrait
	91	skeleton
	91	spacesuit hologram
	91	talking watch
	91	Tokyo Disneyland Sports Festival
Mickey Mouse and Pluto	90	with red heart
Minnie Mouse	80s	made in France
	90	Pedre/as 'Marilyn'
Minnie Mouse and Doll	90	Walt Disney World Doll Convention
Muffy vander Bear	90	bear show mascot
Oliver and Company	80s	Hong Kong/from movie
Paddington Bear	90	Limited edition/Hope
Petey	86	Little Rascals/plastic molded/fliptop

Pink Panther	90	Armitron
	91	black plastic case
Pokey	86	bendable/plastic/LCD
Popeye	86	plastic/fliptop/LCD
	91	Armitron/3-D
	91	Armitron/large
Porky Pig	90s	Armitron
Princess Toadstool	90	plastic/fliptop/LCD
Punisher	91	plastic molded body/LCD
	91	plastic molded face/LCD
Rescuers--Bernard	91	plastic molded/fliptop
Rescuers--Bianca	91	plastic molded/fliptop
Rescuers--Jake	91	plastic molded/fliptop
Roadrunner	80s	Armitron
	90	Sutton Time/plastic case/ graphic band
Robo Cop	90s	plastic molded/LCD
Rocketeer	91	Disney Channel promo
	91	Fossil/checkerboard/running
	91	Fossil/gold-face/flying
	91	Fossil/limited edition
	91	Fossil/symbol
	91	Fossil/white face/flying
	91	Hope/blue metal/LCD
Roger Rabbit	87	Amblin
	87	Amblin/bullet hole
	87	Amblin/silhouette
	87	Armitron
	87	behind bars
	91	digital/running
Schroeder	90	with piano band
Scooby Doo	91	plastic/molded
Scrooge McDuck	90	gold nugget
Simpsons	90	Bart Avenger/fold-over/LCD
	90	Bart/"Get out of my face"
	90	Bart/"No Way"
	90	Bart/"Yo, man"
	90	Bart/Father Time
	90	Bart/fold-over/LCD
	90	family
Snoopy	80s	Armitron
	90	Armitron/gold embossed
	90	Joe Cool
	90	Lucy/moving eyes
	90	Top Secret
	91	moving eyes
	91	TV pictures
Snow White	88	Lorus/50th anniversary
	91	limited edition/revolving seven dwarfs
Spiderman	90	plastic molded body/fliptop/ LCD
	90	plastic molded face/fliptop/LCD
Star Wars	91	Yoda/3-D/hologram
Super Mario	90	plastic/fliptop/LCD
Superman	80s	50th anniversary/Armour meat
	86	plastic molded/LCD/with Lois Lane
	86	Smithsonian
Sylvester	86	plastic/fliptop/LCD
	90	Sutton Time/plastic case/ graphic band
Sylvester and Tweety	90s	Armitron
	90	fliptop LCD
Tale Spin--Baloo	91	plastic molded/fliptop
Tale Spin-- Kit Cloud Kicker	91	plastic molded/fliptop
Tale Spin-- Molly Cunningham	91	plastic molded/fliptop
Tale Spin--Wildcat	91	plastic molded/fliptop
Tasmanian Devil	80s	red plastic/digital
Teenage Mutant Ninja Turtles	91	all 4 in color/gold face
	91	gold-face/limited edition
	91	Raphael/talking
	91	sports model
	91	sports model/Toy Fair
	90	plastic molded/fliptop Turtles-- April O'Neill
Teenage Mutant Ninja Turtles--Don the Undercover Turtle	90	plastic molded/fliptop
Teenage Mutant Ninja Turtles-- Donatello	90	plastic molded/fliptop
Teenage Mutant Ninja Turtles--Donatello Fighter	90	plastic molded/fliptop
Teenage Mutant Ninja Turtles--Leo the Sewer Samurai	90	plastic molded/fliptop
Teenage Mutant Ninja Turtles-- Leonardo	90	plastic molded/fliptop
Teenage Mutant Ninja Turtles-- Leonardo Fighter	90	plastic molded/fliptop
Teenage Mutant Ninja Turtles-- Michaelangelo	90	plastic molded/fliptop
Teenage Mutant Ninja Turtles--Mic-haelangelo Fighter	90	plastic molded/fliptop
Teenage Mutant Ninja Turtles--Mike the Sewer Surfer	90	plastic molded/fliptop
Teenage Mutant Ninja Turtles--Ralph the Space Cadet	90	plastic molded/fliptop
Teenage Mutant Ninja Turtles-- Raphael	90	plastic molded/fliptop
Teenage Mutant Ninja Turtles-- Raphael Fighter	90	plastic molded/fliptop
Teenage Mutant Ninja Turtles--Shredder	90	plastic molded/fliptop
Teenage Mutant Ninja Turtles--Splinter	90	plastic molded/fliptop
Tigger	91	Fossil
Tiny Toon-- Babs Bunny	91	plastic molded/fliptop
Tiny Toon-- Buster Bunny	91	plastic molded/fliptop
Tiny Toon-- Dizzy Devil	91	plastic molded/fliptop
Tiny Toon-- Plucky Ducky	91	plastic molded/fliptop
Tom and Jerry	91	Armitron/3-D
Tommy Tank Engine	91	plastic molded face
Tweety	86	plastic/fliptop/LCD
	90	baseball/flipover/LCD
	90	Sutton Time/plastic case/ graphic band
Ultraman	80s	Gold Japanese/plastic molded/ LCD
	80s	Japanese/plastic molded/LCD
Uncle Scrooge	91	limited edition/gold face
Winnie the Pooh	91	circular disc
Winnie Woodpecker	90	orange plastic
Wizard of Oz	89	for Macy's
Woody Woodpecker	90	50th anniversary
	90	curvex/HaHa
	90	large/round
	90	white plastic
	91	50th anniversary/black/2 heads
Yogi Bear	91	plastic/molded
	91	Ranger chasing Yogi
Yosemite Sam	86	plastic molded/LCD
	90s	Armitron
Zelda	90	plastic/fliptop/LCD
Zoo Crew-- Cool Bunny	91	plastic molded/fliptop
Zoo Crew-- Cool Duck	91	plastic molded/fliptop
Zoo Crew-- Cool Kuala	91	plastic molded/fliptop

Chapter Seven
Advertising Watches

One of the related areas to comic character watches are the advertising watches, sometimes called promotionals or premiums. Tony the Tiger, Charlie Tuna, and Ernie Keebler are cartoon characters used as promotional "spokesmen" for their individual products. In this area, new watches are constantly being issued, and while they are not limited editions, they are issued in limited quantities. Look through the Sunday paper coupon section, save three proofs of purchase, send $3.95, and you too can begin your collection of advertising watches.

There were earlier uses of watches in advertising. The Buster Brown pocket watch was made in 1908. The first advertising character wristwatch was Toppie, used by Tip Top Bread in the mid-1950s. But this type of promotional campaign grew in the early 1970s with Mr. Peanut, Bud Man, Tony the Tiger, and Icee Bear. Charlie Tuna was first issued in 1971 and repeated in 1973, 1977, and 1986. Some of these watches have reached a value of hundreds of dollars, with the most infamous being the Ritz Cracker, followed by the Ralston Tom Mix watch. Even though there are innumerable watches of this type, I will feature only 100 of them in this chapter. By the time this book is published, I will have clipped many more Sunday coupons and received many more promotional watches that will be eligible for my top one hundred.

Bud Man, 1971 Budweiser promotional. Size M.

Bob's Big Boy, 1970 promotional. Size M.

Budweiser Beer, 1971 Budweiser promotional. Size M.

Chocolate Chip, 1971 Nestle promotional. Size M.

Charlie the Tuna. The 1971, 1973, 1977, and the 25th Anniversary 1986 Starkist Charlie the Tuna watches. Size M.

Churchies Chicken, 1971 promotional. Size M.

Disney Channel. Promotional for the Disney Channel. Size M.

Cookie Crisp, 1971 promotional. Size M.

Freezie the Penguin, 1971 promotional. Size M.

Hamm's Beer, 1971 promotional. Size M.

Hallmark Cards. Series made by Picco as an advertisement in the 70s. Size M.

Heinz Ketchup, 1971 promotional. Size M.

Herself the Elf, 1971 promotional for Hallmark cards. Size M.

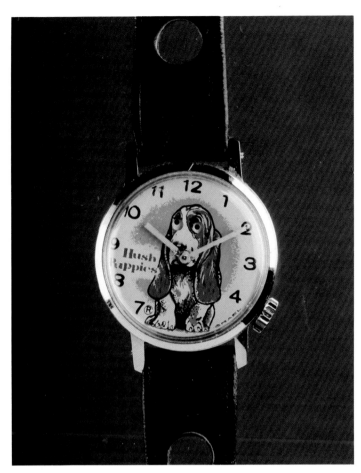

Hush Puppies, 1971 promotional. Size M.

Hughes Aircraft, 1970, Love watch. Size M.

Icee Bear, 1970, by Hasis. Size M.

Irish Spring, 1972 promotional. Size M.

Jack in the Box, 1970 promotional. Size M.

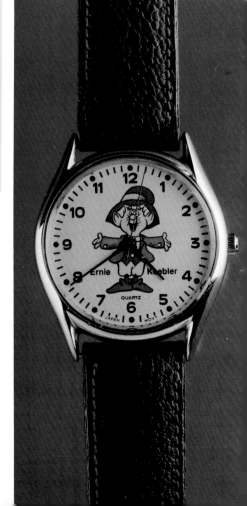

Keebler Elf, 1971 Keebler promotional. Size M.

Little Hans, 1970 promotional for Nestle. Size M.

M & M's, 1989 birthday promotional. Size M.

Major Moonstone, 1970 cereal promotional. Size M.

Mystery Sleuth watch was sold as a fundraiser for Multiple Sclerosis Society Read-a-Thon in the mid-70s. Size M.

Morris the Cat, 1980 promotional. Size M.

Pig, 1972. Distributed by police departments, it stands for pride, integrity, and guts. Size M.

Pillsbury Dough Boy, 1987 promotional. Size M.

Planter's Peanuts, 1971 promotional. Size M.

Planter's Peanuts, 1970 promotional. Size M.

Punchy, 1971 Hawaiian Punch promotional. Size M.

Ritz Cracker, 1971 promotional. Because of the beauty of this watch it is considered to be the #1 collectible promotional watch. Size M.

Reddy Kilowatt. Made in 1972 as a promotional for an electric company. Size M.

Ronald McDonald, 1971 promotional. Size M.

Shirt Tales, 1981, promotional for Hallmark cards. Size S.

Sugar Bear, 1980 promotional. Size M.

Snap Crackle and Pop, Rice Krispies promotional, 1991. Size M.

Super Pan, 1971 promotional for Oster. Size M.

Swiss Miss, 1981 promotional. Size M.

Tom Mix, 1983, for Ralston Purina Corp. Size M.

Tony the Tiger, 1976 Kelloggs promotional. Size M.

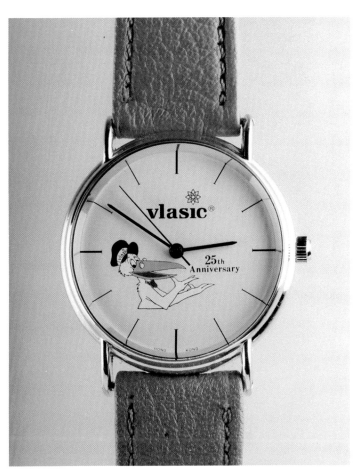

Vlasic Pickle, 1989. 25th Anniversary promotional. Size M.

Toppie, made by Ingraham in 1951 as a promotional for Tip Top bread. **Toppie** qualifies as the first advertising wristwatch. Size S.

W.C. Fritos, 1970 W.C. Fields-like promotional for Fritos. Size M.

ADVERTISING WATCHES

WRISTWATCHES

NAME	YR	NOTES
Alvin	91	Del Monte
Apple	68	Beattles apple
Bartlett and James	80s	yellow and red/Gallo promo
Bob's Big Boy	70	animated hands
Bud Man	71	animated hands
Budweiser	72	Jay Ward
	91	bottle watch
California Raisins	88	plastic molded/fliptop
Campbell Soup Kids	80s	boy/short pants/small
	80s	girl/green dress/large
	80s	girl/yellow dress/small
	80s	male/blue jeans/large
Captain Moonstone	80s	brown/silver/promo for cereal
Care Bear	83	Bradley/promo for American Greetings
Charlie the Tuna	71	Charlie on back by #3
	73	Charlie on back by #1
	77	Charlie on back by #9
	86	25th anniversary
Chocolate Chip	82	Chocolate Chip Cookie
Churchie	71	Helbros/promo for chicken
Cookie Crisp	71	promo
Cracker Jack	89	box
Crayola	80s	fliptop/LCD
Ernie Keebler	71	Keebler elf
Freezie Penguin	71	large/silver
Green Giant	80s	promo
Hallmark	71	Happy Days/Picco/blue flowers/case and band
	71	Happy Days/pink flowers/band and case
	71	Herself the Elf/Picco/pink plastic
Hallmark	76	girl and flowers/yellow plastic band and case
	76	Take Time to Care/girl and dog
	78	Herself the Elf/promo
Hamm's Beer	71	Calendar
Heinz Ketshup	71	moving bottle
Hot Wheels	70	red & green car on plastic disc
	70	red & green car on second hand
	70	rotating bezzle
Hughes Aircraft	70s	Love is../naked boy and girl
Hush Puppies	70s	Israel
Icee Bear	71	17-jewel Swiss/Hasis
	75	Bradley
Irish Spring	72	Shamrock
Jack in the Box	70s	red dial
Lemon Drop Kids	70s	Swiss analog
Lemon Frog	80s	yellow frog/promo
Little Hans	70	Nestles promo
M & M	87	yellow case
Monopoly	86	white face/black band
Morris the Cat	80s	9-lives
Mystery Sleuth	70s	Multiple Sclerosis Read-a-Thon
Oster	71	blue background/promo
Pacman	80	Bradley
Pepsi	91	Pepsi bottle/digital
Pillsbury Doughboy	87	blue plastic
Planters Peanut	70	yellow disc
	71	blue digital
Punchy	70	Hawaiian Punch
Raid	75	17-jewel
Reddy Kilowatt	72	yellow face/promo
Ritz Cracker	71	made for Ritz
Ronald McDonald	70s	promo
Scrubbing Bubbles	75	promo
Shirt Tales	81	for Hallmark
Simpsons	91	Butterfinger/promo
Snap-Krackle-Pop	91	Rice Krispies promo
Snickers	90	revolving disc/promo
Southwest Airlines	90	Fly Shamu
	90	"I love..."/rotating disc
Splash Mountain	89	grand opening/for Disneyland
	89	water/for Disneyland
Sugar Bear	80s	moving eyes
Swiss Miss	81	standing figure
Teddy Grahams	91	hologram/promo
Three Musketeers	89	gray and red
Tom Mix	83	made for Ralston
Tony the Tiger	76	for Kelloggs
	80	Kellogg/set of 4 characters
UAL Plane	90	promo
Vlasic Pickle	89	25th anniversary
W. C. Fritos	70s	charicature of W.C.Fields
Welch's	89	purple/promo
Where's the Beef	84	fliptop/for Wendy's
Ziggy	78	American Greetings cards

Chapter Eight
Political Watches

Political cartoons have been part of our lives for centuries. However, it was not until the early 1970s that this type of commentary art form was placed in a watch and sold to the public. These watches were produced by privates enterprises and sold for profit, not issued by political fund-raising committees.

The political commentary displayed on many of the watches not only make them interesting but portray a part of our history. Some of my favorites include: Sam Erwin Watergate, where each number is the name of a Watergate conspirator; Richard Nixon, with both hands raised in the "V" for victory sign and saying "I'm not a crook," while his eyes shift at the tick of the watch; Lester Maddox, with a baseball bat in one hand and a chicken drumstick in the other; George Wallace, as the "Fighting Judge"; Gerald Ford, wearing a football helmet while on skis; Ronald Reagan, riding an elephant; Jimmy Carter, with the body of a peanut; Dan Quayle, where the numbers are completely out of order; George Bush, where the numbers are backwards; and the Ted Kennedy water-proof, where he is under water. New ones are issued as time passes.

An interesting concept pursued by Comic Time is their limited edition "political superstar" watches, issued for each major election year. They feature six Democrats with numbers of the left, and six Republicans with numbers on the right.

While the majority of political watches are issued for profit, there are those that have been issued by the candidate's campaign staff as fund-raisers. They are normally as innovative as "Ford in '76". Of the fifty watches I feature in this chapter, you may note that I have not included many of these.

Uncle Sam, 1972, Swiss made. Size M.

Democratic Donkey, 1971, Dirty Time Co. Size M.

Democratic Donkey, 1972, Swiss made. Size M.

GOP Elephant, 1972, Swiss made. SIze M.

GOP Elephant, 1971, Dirty Time Co. Size M.

Spiro Agnew, 1971, All American. Size M.

Spiro Agnew, 1971, Dirty Time. Golf and tennis hands. Size M.

Spiro Agnew, 1971, Dirty Time. Peace hands. Size M.

Spiro Agnew, 1971, Sheffield. Size M.

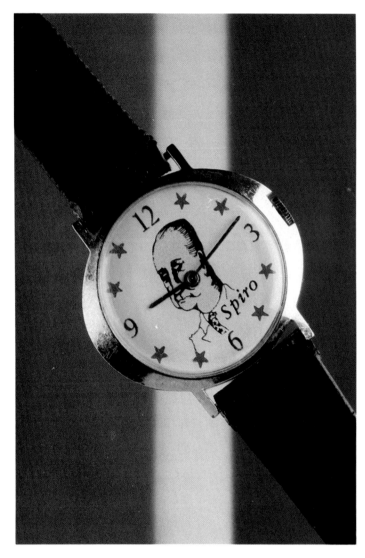

Spiro Agnew, 1971. Face. Size M.

Spiro Agnew, 1970. Body. Size M.

Barbara Bush. Part of a series that includes George Bush, Gorbachev, Brian Mulroney, and Princess Di. Size M.

George Bush. Part of a political series that included Princess Di, Brian Mulroney, Barbara Bush, and Gorbachev. Size M.

George Bush, 1990. Part of a series that included Dan Quayle, and Gorbachev. Notice that Bush's numbers are backwards. The illustrator's attitude about our President are obvious. See Dan Quayle watch for his opinion of our Vice President. Size M.

Bush and Bush Lite, 1991. Depiction of George Bush and and Dan Quayle leave little doubt of the artist's feelings about Dan Quayle. Part of a series of five. Size M.

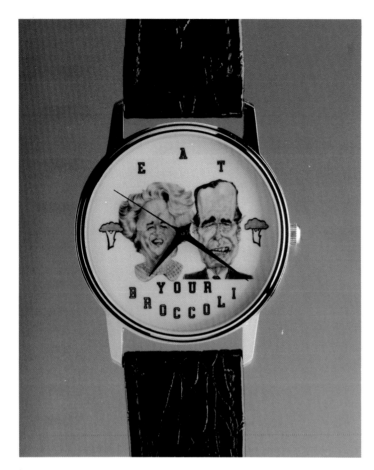

George and Barbara Bush, 1991, called "Eat Your Broccoli." Part of a series of five. Size M.

Bush and Gorbachav. "Peace Summit, 1990. Part of a series of five. Size M.

Bush and Gorbachev, 1990. A commemorative edition given to athletes in Seattle for the World Games. Size M.

Jimmy Carter, 1976, Goober Time. Size M.

Jimmy Carter, 1980, Timely Creations. Size M.

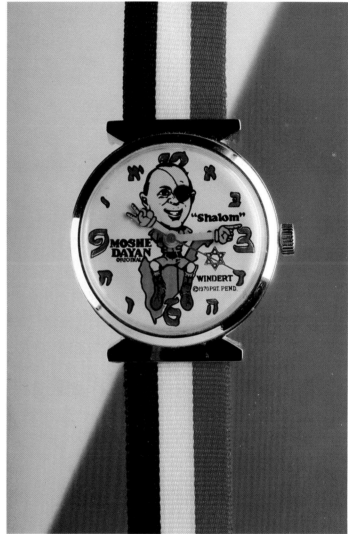

Moshe Dayan, 1970, Windert. Size M.

Fidel Castro, 1991. Part of a series of five. Size M.

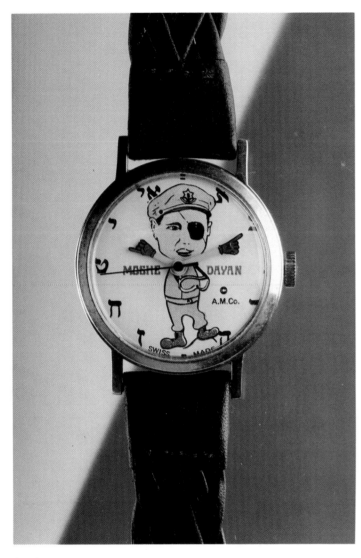

Moshe Dayan, 1970, AMC. Size M.

Princess Di and Prince Charles, 1990. Part of a series of five that included Brian Mulroney, George Bush, Barbara Bush, and Gorbachev. Size M.

Sam Erwin, 1972, Honest Time. Size M.

Gerald Ford, About Time, 1980. Size M.

Gerald Ford, 1976 fundraiser for the election committee. Size M.

Gerald Ford. Promotional for Ford in 1976. Size M.

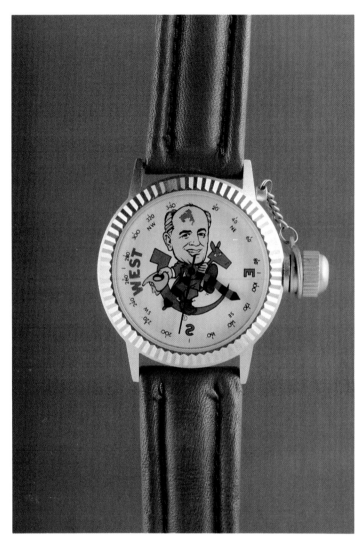

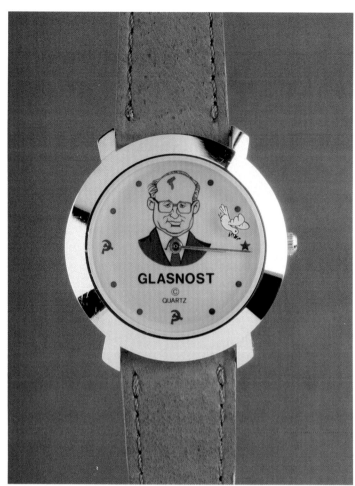

Gorbachev. Part of a series that included George Bush, Barbara Bush, Brian Mulroney, and Princess Di. Size M.

Gorbachev, 1990. Part of a series that included Dan Quayle, George Bush. He's called the "Ukranian Cowboy." Size M.

Gorbechev, 1991. Part of a series of five. Size M.

Hippie, made in the early 70s by Windart. Size M.

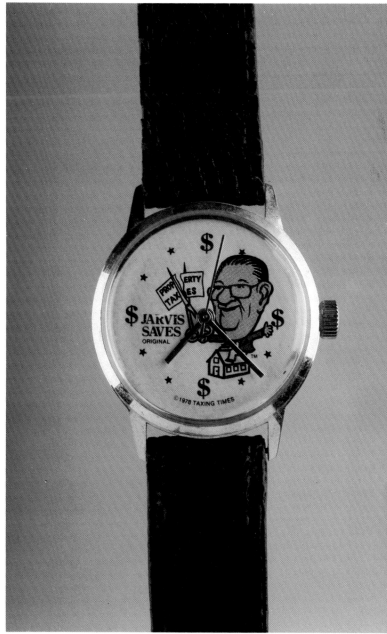

Howard Jarvis, 1978. Used as a fundraiser. Size M.

Ted Kennedy, 1970, Peace Time. Size M.

Teddy Kennedy, 1971, waterproof. Size M.

Henry Kissinger, 1972, Timely Creations. Size M.

Henry Kissinger, 1974, Trying Time. Size M.

Lester Maddox, 1971, by ATC. Size M.

John Lindsey, 1971, Dirty Time Co. Size M.

Brian Mulroney. Part of a series of five that included Princess Di, George Bush, Barbara Bush and Gorbachev. Size M.

Lester Maddox, 1971, autographed. Size M.

Dickey Nixon, 1971, Dirty Time Co. Size M.

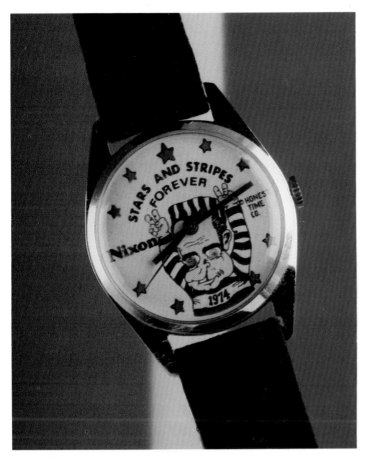

Dick Nixon, 1974, Honest Time. The eyes move on this watch. Size M.

Dick Nixon, 1974, Tricky Tock. Size M.

Dick Nixon, 1974, All American Time. The eyes move on this watch. Size M.

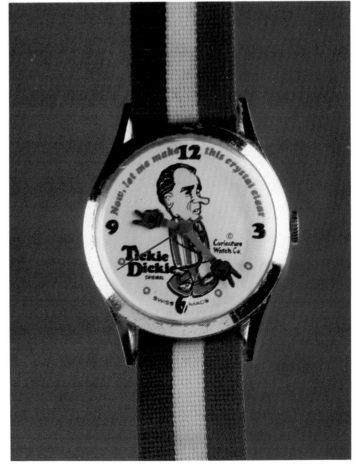

Tickie Dickie, 1974, Caricature Watch. Size M.

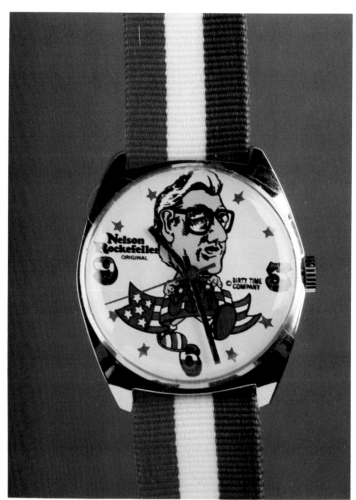

Nelson Rockefeller, 1971, Dirty Time Co. Size M.

Dan Quayle, 1990. Part of a series that included George Bush and Gorbachev. Notice the numbers are all mixed up. The illustrator's attitude towards our Vice President is obvious. See George Bush for his opinion of our President. Size M.

Ronald Reagan, 1980, Timely Creations. Size M.

Ronald Reagan, Fraxe, 1970. Made prior to his presidency. Size M.

George Wallace, Bill Dinken, 1980. Size M.

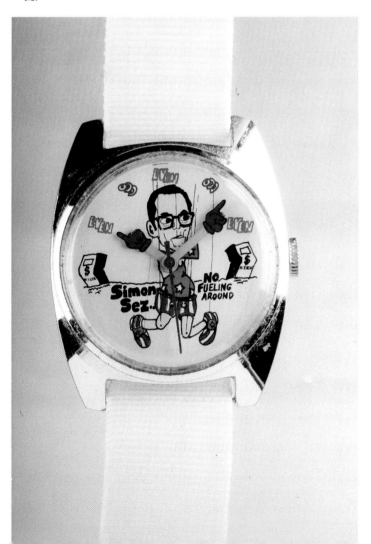

Simon Says, 1970s. This watch, featuring William Simon, obviously was made to comment on our fuel crisis. Size M.

POLITICAL WATCHES

WRISTWATCHES

NAME	YR	NOTES
Amy Carter	70s	Timely
Barbara Bush	90	with Millie/First Lady
Brian Mulroney	90	dollar signs/Canada Prime Minister
Bush and Gorbachev	90	Goodwill Games
Castro	90	Cuba
Dan Quayle	89	Krazy Time/backward numbers
Democrat Donkey	71	Dirty Time
	71	Swiss date
George Bush	89	Krazy Time/pink slippers
	90	"Read my Lips"
	90	Bush and Bush Lite
	90	Gorbachev summit
	90	with Barbara/"Eat your broccoli"
George Wallace	70	fighting judge
Gerald Ford	70s	in car
	76	"Love is Ford"
	76	Foot in Mouth
	76	skis and helmet
GOP Elephant	71	Dirty Time
	71	Swiss date
Gorbachev	90	Krazy Time/Ukranian Cowboy
	90	Perestroika
	90	revolving dove
Henry Kissinger	72	super-kiss
	74	in superman outfit
Howard Jarvis	78	Jarvis Saves
Jimmy Carter	76	peanut body
	80	riding donkey
John Anderson	70s	For President
John Lindsey	71	Dirty Time
Lester Maddox	71	bat and drumstick
	71	riding bike
Moshe Dayan	70	eyepatch
	70	uniform
Nelson Rockefeller	71	Dirty Time
Nixon	70s	I'm Not a Crook
	70s	Nixon steady
	70s	Tricky Tick
	71	Dicky/Dirty Time
	74	moving eyes/Not a Crook
	74	Stars and Stripes Forever
	74	Ticky-Dicky
	74	Watergate Bug
Pig	70	pig policeman
Princess Di	90	with Prince Charles
Ronald Reagan	70	cowboy
	80	riding elephant
Saddam Hussein	91	color/moving eyes
Sam Erwin	74	Watergate
Simon Says	70s	"No Fueling Around"
Spirit of 76	76	3 flag holders
Spiro Agnew	71	Dirty Time/peace hands
	71	Dirty Time/tennis and golf
	71	face
	71	peace symbol
	71	Sheffield
	71	Striped tails
Super Jew	70	Jewish superman/rabbi in phone booth
Ted Kennedy	70	peace-hands
	71	Water-proof
Uncle Sam	71	Swiss date

Chapter Nine
Personality Watches

Personality watches first pictured sports heroes, such as Dizzy Dean and Babe Ruth, both issued prior to 1950. Eighteen years passed before this type was introduced again in 1968 and brought us the Beatles' "Yellow Submarine", Bob Hope, Arthur Fiedler, W. C. Fields, Laurel and Hardy, and later Evil Knieval, Jerry Lewis, Jackie Gleason, and Lucille Ball.

In 1985 when Bradley Watch Company was going out of business, they issued "The Oldies" series as a last effort to generate sales. Personalities included Abbott and Costello, Laurel and Hardy, W. C. Fields, Elvis Presley, Marilyn Monroe, Emmett Kelly, Charlie Chaplin, and the Three Stooges. They appeared in various types of cases and packaged in blister-pack featuring graphics of each character. This series makes a delightful display of nostalgia. Bradley's last watch was a backwards Charlie Chaplin, never issued for sale to the public, which became available in late 1988. It is a must for everyone's collection.

A new personality watch has become extremely desirable—the plastic molded PeeWee Herman. This watch originally sold for $3.00 in early September, 1991, but is now being offered for $75. The Bradley Muhammed Ali watch is being speculated by many dealers who believe that someday soon it will become more valuable than it is today.

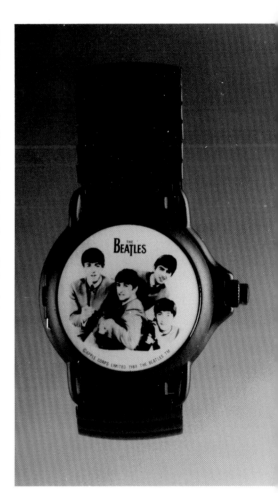

Beatles, 1991. Part of a series of many stars done in this manner. Size M.

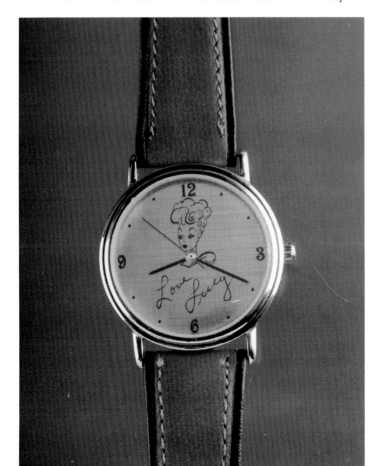

Muhammed Ali, *see page 163.*

Lucille Ball, distributed in 1986 in conjunction with her last television show. Size M.

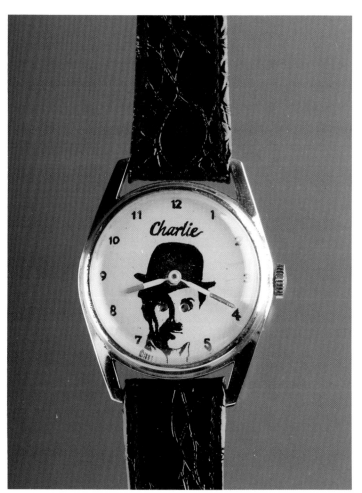

Charlie Chaplin, 1975, with moving eyes. Same series as Groucho Marx. Size M.

Beethoven, 1971, by Offbeat Time Co. Size M.

Charlie Chaplin, *see also page 130.*

Charlie Chaplin, 1972, made by Cadeaux. Animated cane. Size M.

Dizzy Dean, 1935. *Courtesy of Jeff Cohen.*

Dizzy Dean, 1935, New Haven. The first personality pocket watch. *Courtesy of Roy Ehrhardt.*

Dizzy Dean, 1935, Made by Everbrite. The first baseball player watch. Size M.

W.C. Fields. Made in 1971 for Jay Ward Productions. This was one of his favorite characters, as was Buster Keaton. Size M.

Arthur Fiedler. A 1971 Commemorative to Mr. Pops. Size M.

W.C. Fields, 1971, Dirty Time. Size M.

W.C. Fields. Part of the 1985, "Oldies" series made by Bradley that included Charlie Chaplin, the Three Stooges, Laurel and Hardy, Elvis Presley, Marilyn Monroe, Abbott and Costello, and Emmett Kelly Jr. The watches came in two sizes and three different bands.

Bob Hope. A commemorative of the 1971 USO tour. Size M.

Jackie Gleason, 1985. Issued by Showtime to commemorate the showing of the lost episodes of the Honeymooners. Size M.

Buster Keaton, 1971. Made for Jay Ward Production. This was one of his favorites, as was W.C. Fields. Size M.

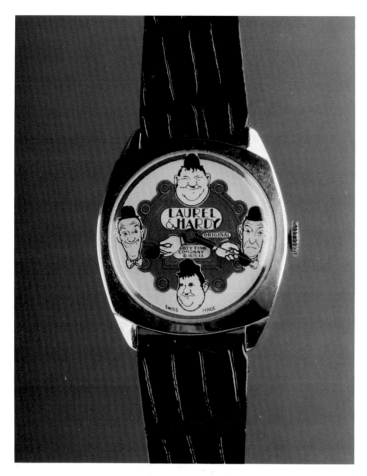

Laurel and Hardy, 1971, made by Dirty Time. Size M.

Jerry Lewis. A 1971 watch given to people working on the MD telethon. Size M.

Rod Laver, 1970, to commemorate his playing at Wimbledon. Size M.

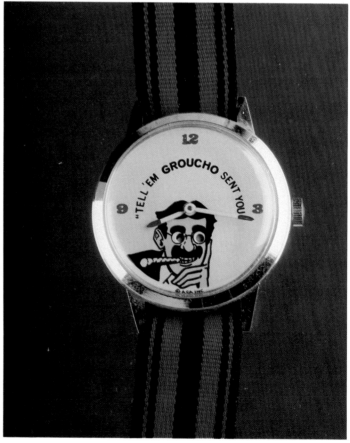

Groucho Marx, 1975. Part of the series that included Charlie Chaplin. Size M.

Babe Ruth, made by Exacta, 1949. Came with metal expansion band and was packaged in a plastic baseball. Size M.

Three Stooges, 1991 reissue. This is part of a series of reissued watches made by Sheffield. Size M.

Three Stooges, *see also page 180.*

Babe Ruth, 1949, made by Exacta Time. In order for the Babe Ruth to be authentic the numbers and the hands must shine in the dark. Size M. *Courtesy of Jack Feldman.*

PERSONALITY WATCHES

WRISTWATCHES

NAME	YR	NOTES
Abbott and Costello	85	Bradley/oldies
Arthur Fiedler	71	Mr. Pops
Babe Ruth	49	Exacta Time
Beatles	68	yellow submarine with band
Beethoven	70	Off-beat Time
Bob Hope	71	USO tour
Buster Keaton	72	Jay Ward
Charlie Chaplin	72	black and white
	75	moving eyes
	85	Bradley/oldies/backwards
	85	Bradley/oldies/large/gold
Dizzy Dean	35	with metal band
Elvis Presley	85	Bradley/oldies
Emmett Kelly	85	Bradley/oldies
Evil Knieval	76	Bradley/white face
Freud	91	50-minute hour
Groucho Marx	70s	moving eyes
Harlem Globetrotters	70s	CBS
Jackie Gleason	85	Showtime
Jay Ward	71	Jay Ward/caricature
Jerry Lewis	71	red head
Lassie	68	small size
Laurel and Hardy	71	Dirty Time
	85	Bradley/oldies/small gold
Lucille Ball	86	line drawing
Marilyn Monroe	85	Bradley/oldies
Mr. T	70s	Bradley
Muhammad Ali	70s	Bradley
Muhammed Ali	80	I'm the Greatest
Peewee Herman	88	plastic molded/LCD
Rod Laver	70	one hand tennis ball/other, racket
Shirley Temple	54	name only
Three Stooges	85	Bradley/oldies/small white plastic
W. C. Fields	71	Dirty Time
	71	Jay Ward/with saying
	72	Jay Ward
	85	Bradley/oldies/large black plastic
Whoopie Goldberg	88	Jumpin' Jack Flash

Appendix A: WRISTWATCHES
1933-1972

Alice in Wonderland	50	blue backgrnd/silver case
	54	pink case and band
	58	coming out of flower
	58	Timex/name only
	72	animated Madhatter
All-Star Baseball	66	black dial
	66	green dial
All-Star Football	66	autograph
Andy Panda	60s	blue plastic/Japanese
	71	red face
	72	small size
Annie Oakley	51	moving gun
Ape--George of the Jungle	71	Jay Ward
Archie	72	Rouan/revolving disc
Astro Boy	60s	metal/Japanese
Ballerina	56	Bradley/animated feet
	62	Bradley/hands at 1 and 10
	68	Bradley/hands at 4
BamBam	71	Prince Roable/small size
Bambi	48	birthday
	49	luminous
Barbie	63	facing left
	64	with Ken
	71	facing right/blue rim
Baseball Player	60	animated reflector
Batman	66	plastic by Gilbert
	71	Dirty Time
Big Bad Wolf	34	red with band
Blondie	49	Dagwood & animals
Bongo	48	birthday/large figure
	48	birthday/small figure
Boris	71	Jay Ward
Bozo the Clown	70	Capitol Record
Brick Bradford	71	Hi-Time
Bronco	66	Gilbert/plastic
Bronco Rider	60	animated reflector
Brutus	60s	metal/animated hands/ Japanese
Buck Rogers	71	Huckleberry Time/ pocket wristwatch
Buffy and Jody	71	Sheffield/Time Tell
Bugs Bunny	51	bush in ctr/carrot hands
	51	carrot ctr/green #/ hands

	51	no ctr/org #/blade hands
	72	animated hands/red bow-tie
	72	Bugs about baseball
	72	face and hand only/ small size
	72	face only/small size
	72	Lafayette
Bullwinkle	71	Jay Ward
Buster Brown	71	Buster & Tige
Captain Liberty	54	embossed airplanes on band
Captain Marvel	48	pink gold
	48	round silver
Capt. Marvel, Jr.	48	tonneau
Casper	60s	pink face
Cat in the Hat	72	see-through back
Chitty Chitty Bang Bang	71	Sheffield/from movie
Cinderella	50	pink backgrnd/blue plastic case
	50	pink backgrnd/metal case
Cinderella	58	castle at 12
	58	Timex/name only
	60s	European
Cool Cat	71	Sheraton
Cowboy	69	moving gun
Cowgirl	71	Joan Walsh Anglund
Daffy Duck	71	Sheffield
	71	Sheraton
Daisy Duck	47	rectangular
	48	birthday/large figure
	48	birthday/small figure
	49	luminous
	49	large size
	49	rectangular/standing
Dale Evans	57	blue tonneau
	57	gold tonneau
	57	pink tonneau
	57	rectangular
	57	silver tonneau
	60	flasher
	62	horseshoe/white
Daniel Boone	69	Powderhorn
Danny & the Black Lamb	47	red rectangular
Davy Crockett	51	moving knife/Muros
	54	round silver

	54	small, round/green plastic
	54	with gun/lettering by Liberty
	56	standing yellow/ rectangular
	56	tonneau/yellow
Davy Crockett	64	English
Deputy Sheriff	60s	small size
Dick Tracy	34	large size
	48	rectangular
	48	tonneau
	51	moving gun
Donald Duck	35	Mickey disc/leather band w/Mickeys
	47	blue face/round/no circle
	47	blue face/rectangular silver case
	47	rectangular gold
	48	birthday/large figure
	48	birthday/small figure
	49	luminous
	50	round/silver/small
	54	Swiss made
	58	Timex/name only
	60s	Phinney Walker
Dopey	48	birthday/large figure
	48	birthday/small figure
	60s	white clear plastic/ Japanese
Draemon	70s	Japanese cat
Drooper	71	large size
Dudley Do-right	71	Jay Ward
	71	Jay Ward/hand-painted
Elmer Fudd	71	Sheraton
Felix the Cat	60s	Japan/black & white silhouette
	71	Sheffield
Flash Gordon	71	Precision Time
Fred Flintstone	71	Prince Roable/ani- mated hands
G.I. Joe	66	Gilbert
Gene Autry	48	rectangular/on Cham- pion
	48	round face
	51	moving gun
George of the Jungle	71	Jay Ward
Girl from U.N..C.L.E.	66	pink dial
Goofy	71	backwards/Helbros
Hippie	70	Peace Time
Hoky Poky	49	moving card hand
Hopalong Cassidy	60	Great Britain/small/ metal/different collar
	50	large
	50	small, black plastic

Hoppity Hoop	71	Jay Ward
Howdy Doody	54	large/moving eyes
	54	tonneau w/friends
	71	animated hands
Huckleberry Hound	66	Bradley
Humphrey T. Bear	60s	Q and Q
Humpty Dumpty	67	blue metal face
James Bond 007	66	Gilbert
Jerry (of Tom and Jerry)	60s	white clear plastic/ Japanese
Jiminy Cricket	48	birthday/large figure
	48	birthday/small figure
	49	luminous
Joe Carioca	48	birthday/large figure
	48	birthday/small figure
	53	animated hands
Joe Palooka	47	small/rectangular
Junior League	56	Bradley
Junior Nurse	56	Bradley
L'il Abner	50s	small/silver/black & white face
	51	moving flag/front view
	51	moving flag/side view/ salute
	51	moving mule
Little Annie Fanny	70	Playboy bunny
Little King	71	O. Soglow
Little Nell	71	Jay Ward
Little Pig	47	with fiddle
Little Red Riding Hood	60s	animated wolf
Lone Ranger	39	large size
	47	rectangular
	51	round Silver
Louie Duck	47	rectangular
Lucy	58	Bradley/writing under 6
Majorette	60	animated reflector
	69	Bradley
Man from U.N.C.L.E.	66	blue dial
Mary Marvel	48	small size
Merlin the Magic Mouse	71	Sheffield
Mickey Mouse	33	English w/beard
	33	gold case & band/ copper hands
	33	with metal band
	38	disc/women bracelet
	39	gold/blade
	39	silver/blade
	46	Mickey Kelton/head only/gold face
	46	Mickey Kelton/head only/white face
	47	gold rectangular
	47	round/white-face
	47	silver rectangular
	48	birthday series/small figure

	48	birthday/large figure
	49	luminous
	50	round/silver/small
	50	round/silver/small/ Ingersoll under
	58	red plastic
	58	Timex/name only
	60s	blue plastic/baseball player/Japanese
	60s	moving eyes/made in Israel
	68	Timex
	69	Phinney Walker
	71	Elgin electric
	71	Helbros
	71	Helbros electric
	71	Helbros/day/date
	71	Timex electric
	71	Vantage/Disneyland/ clear back
	71	Vantage/Disneyland/ white hands
Minnie Mouse	58	round/silver
	71	Helbros
	71	Timex
Mush Mouse	60s	metal/animated hands/ Japanese
Natasha	71	Jay Ward
Olive Oil	60s	metal/animated hands/ Japanese
Oliver	71	Sheffield/from musical
Orphan Annie	34	large size
Orphan Annie	47	small/rectangular
	48	tonneau
	71	Hi Time/large size
Peace Mouse	70	Peacetime Company
Pebbles	71	Prince Roable/small size
Peter Pan	72	Rouan/revolving disc
Pinnochio	48	birthday/large figure
	48	birthday/small figure
	49	luminous
	60s	small size/on strings
	66	Bradley/lying on side
Pluto	48	birthday
	49	luminous
Popeye	35	with friends
	48	unknown/friends on face
	49	hexagonal case
	60s	metal/animated hands/ Japanese
	66	Bradley
	71	large, Sheffield/ animated arms

Porky Pig	49	rectanglular
	49	round
	71	Sheffield
	71	Sheraton
Punkin Head	48	Ingraham/rectangular
Puss 'n Boots	47	Saro
Quick Draw McGraw	60s	green plastic/Japanese
	60s	metal/animated hands/ Japanese
	66	Bradley
	71	small size
Raggedy Ann		
Red Rider	49	round
Roadrunner	71	revolving disc/with Wilee Coyote
	71	Sheffield
Robin Hood	56	rectangular
	58	round
Rocky	71	Jay Ward
Rocky and Bullwinkle	60s	metal/Japanese
Rocky Jones	54	rectangular
Roy Rogers	51	on Trigger
	54	sitting on Trigger/ facing right
	55	large face/yellow
	57	rectangular/green
	57	tonneau/green
	60	flasher
	60	reflector/running horse/inscribed case
	62	white
	65	on Trigger
Rudolph the Red Nosed Reindeer	47	Ingersoll
Scooby Doo	70	yellow-dial/animated hand
	71	21-jewels/with date
See-saw Marjorie Daw	54	small size
Shep	71	Jay Ward/pink elephant
Skipper	64	Mattel
Smitty	34	large
Smokey Bear	71	Hawthorne
Smokey Stover	71	Hi Time
Smurfs--Schlumpf	60s	including Smurf band
Snidely Whiplash	71	Jay Ward
Snoopy	68	Timex/lying on top of doghouse
	69	black backgrnd/white Snoopy
	69	orange backgrnd/white Snoopy
	69	red backgrnd/white Snoopy
	69	yellow backgrnd/ orange Snoopy
	69	yellow backgrnd/white Snoopy

Name	Year	Description
Snow White	47	rectangular
	48	Ingersoll/round
	50	white backgrnd/silver
	58	Timex/name only
	58	Timex/with Dopey
	58	white backgrnd/yellow plastic
Space Explorer	54	compass in middle
	54	yellow rocket ship
Space Mouse	60s	blue face
Space Patrol	50	small/silver
Spaceman	67	yellow-gold face/Japan
Spooky	60s	blue and black
Super Chicken	71	Jay Ward
Superman	39	large size
	47	rectangular
	48	tonneau
	55	bolt-hands/green face
	62	Bradley/Superman flying
	68	Revolving disc
Sylvester	71	Timesetters
Tammy Tell Time	60s	Teach Time
Terry Tell Time	60s	Teach Time
Texas Ranger	51	Muros
Tom and Jerry	60s	metal Tom w/fork/ Jerry, second-hand/ Japanese
	70	Jerry on second-hand
Tom Corbett	51	tonneau
Tom Mix	35	with embossed band
Top Cat	60s	red plastic/Japanese
	60s	white plastic/Japanese
Topo Gigio	70	big ears
Toppie	51	Ingraham/tonneau
Tweety	60s	metal/animated hands/ Japanese
	71	Timesetters
Underdog	71	animated hands
Wendy	72	Rouan/revolving disc
Westerner	54	Ingraham
Wilee Coyote	71	Sheraton
Wizard of Oz	60s	small size
Woody Woodpecker	50	rectangular
	50	tonneau shaped
	71	moving Woodpecker
	72	plastic bubble
	72	Rouan/revolving disc
Yogi Bear	60s	with BooBoo/brown plastic
	68	Bradley
	71	Prince Roable
Yosemite Sam	71	Timesetters/animated hands
Zorro	50	black face

Appendix B: POCKET WATCHES
1933-1972

Betty Boop	34	embossed back
Big Bad Wolf	34	Ingersoll/embossed/ winking eye
Buck Rogers	35	embossed back/ lightning bolt hands
	71	Huckleberry
Buster Brown	71	Huckleberry
Captain Marvel	48	New Haven
Captain Midnight	48	Ingraham
Dan Dare	53	Ingersoll/double animation
Dick Tracy	48	Ingersoll
Donald Duck	39	Ingersoll/Mickey decal
	54	Swiss-made
Flash Gordon	39	New Haven
	71	Huckleberry
Hopalong Cassidy	50	U.S. Time/black/round
Jeff Arnold	53	Ingersoll/animated pistol
Lone Ranger	39	New Haven/with decal
Mickey and Minnie	71	Love/Al Horen

Mickey Mouse	33	Ingersoll/short-stem/ embossed back
	33	long-stem/Ingergoll/ embossed back
	34	English/fat Mickey
	34	English/rat Mickey
	38	lapel/black/round/with decal
Moon Mullins	33	Ingersoll
Peter Pan	48	Ingraham
Popeye	34	New Haven/with friends
	35	New Haven/no friends
Roy Rogers	59	Bradley/on Trigger
Rudy Nebb	33	Ingraham
Skeezix	36	Ingraham
Smitty	36	New Haven
Superman	60	Bradley
Three Little Pigs	39	prototype
Tom Mix	34	Ingersoll/embossed back

Appendix C: CLOCKS
1933-1972

Bambi	64	Bayard
Batman	69	Bradley
Big Bad Wolf	34	Ingersoll/jaw opens
Bugs Bunny	51	electric clock
	51	Ingraham/alarm
	60s	Seth Thomas
Charlie McCarthy	38	Gilbert/mouth opens
Davy Crockett	54	Hadden/horse bucks
	55	Pendulette
Donald Duck	34	Ingersoll
	50	Glen/moving head
	64	Bayard
Hopalong Cassidy	50	U.S.Time/black/round
Lady	55	Allied/plastic/figure
Mickey Mouse	33	Ingersoll electric/ square/moving head
	33	Ingersoll/wind-up/ square/with disc
	34	art deco/desk model
	34	English/alarm
	34	English/wind-up
	34	Ingersoll/round/alarm/ moving head
	36	Bayard/wall clock
	47	Ingersoll/plastic/alarm/ luminous hands

Pinnochio		
Pluto		
Popeye		
Roy Rogers		
Schmoo		
Sleeping Beauty		
Snow White		
Woody Woodpecker		

47	Ingersoll/round/metal case
49	Ingersoll/plastic/ electric
49	round/thick metal
55	Allied/plastic figure
60s	green case/Phinney Walker
64	Bayard
64	Bayard
53	Allied/plastic/figure
64	Bayard
34	New Haven/round/ friends on outside
68	Smits/animated Sweet-Pea
51	Ingraham/moving horse
47	New Haven/white plastic/figure
60s	round
64	Bayard
50	Columbia/wall clock
50	Columbia/Woody's Cafe

Comic Timepieces

Price Guide

The values listed are for watches in working order and in fine to extra fine condition. The price paid will vary from geographical location, and will be affected by the eagerness of the buyer, the willingness of the seller, and whether the watch is purchased in a retail shop, antique show, or flea market.

The bottom line is that prices can vary greatly, and this guide can only act as a general guide and should not be used to set a price for any particular item. The authors do not claim to be the final authorities on prices and assume no responsibility for financial loss or gain based on the use of this guide.

Pg.	Pos.	Watch	Box
12	TL	1500-2000	500-1500
	TR	1500-2000	500-1500
13	TL	1500-2000	1000-2500
14	TR	1500-2000	1000-1500
	BL	1000-1500	1000-1500
	BR	1000-1500	1000-1500
15	TL	4000-6000	
	BR	3000-5000	
16	TL	1500-2000	1000-2000
	BL	700-900	1000-1500
17	TL	600-800	1000-1500
	BL	1200-1800	1000-1500
	BR	1500-2000	1000-1500
18	TL	1200-1500	1000-1500
	TR	1200-1800	1000-1500
	BL	1200-1800	1000-1500
19	BL	4000-6000	
20	TL	1000-1500	1000-1500
	TR	900-1400	
	BL	1000-1500	1500-2000
	BR	1500-2000	
21	TR	1000-1500	1500-2000
22	BR	1000-1500	1000-1500
23	TR	1500-2000	1500-2000
	BL	1500-2500	1000-2000
24	TL	1000-1500	1000-1500
	TR	2000-2500	1500-2000
	BL	2000-2500	
	BR	1200-2000	1000-1500
25	TL	1500-2500	
	BR	1000-1600	1000-1500
27	B	300-500	700-1500
			(each)
29	BL	600-1000	500-1000
	BR	300-500	500-700
30	BL	700-1200	500-1000
	BR	800-1500	500-1000
31	BL	1000-1500	1000-1500
	TR	700-1200	500-1000
32	TL	700-1200	500-1000
	TR	800-1500	500-1000
	BR	700-1200	500-1000
33	TL	700-1200	500-1000
	BR	1000-1500	500-1000
34	TL	100-400	500-1000
	BL	100-200	300-500
	BR	700-1200	500-1000
35	TL	700-1200	500-1000
	TR	700-1200	500-1000
	BL	300-500	700-1500
36	BL	1000-1500	1000-1500
37	TL	700-1200	500-1000
	TR	800-1500	500-1000
	BL	700-1200	500-1000
38	TL	500-800	400-1000
	TR	600-1000	500-1000
	B	500-800	400-1000
39	TL	1000-1500	1500-2500
	TR	400-700	500-1000
40	TL	400-700	500-1500
	TR	300-700	500-1000
	BR	500-800	500-1000
41	TL	200-500	500-1000
	BL	400-700	500-1500
	TR	800-1200	500-1000
42	TL	700-1200	500-1000
	TR	700-1100	500-1000
	BR	800-1200	500-1000
44	TL	700-1200	500-1000
	TR	800-1500	500-1000
	BR	700-1200	500-1000
45	TL	500-800	500-1000
	TR	800-1500	500-1000
	BL	800-1500	500-1000
46	TL	700-1200	500-1000
	TR	700-1200	500-1000
	BR	800-1500	500-1000
47	TL	500-800	500-1000
	TR	500-800	500-1000
	BL	800-1200	500-1000
48	TL	500-1000	500-1000
	TR	200-600	500-1000
	BL	400-800	500-1000
	BR	600-1200	
49	BR	500-1000	500-1000
50	TL	500-1000	500-1000
51	TR	800-1500	500-1000
	BR	700-1200	500-1000
52	TL	800-1500	500-1000
	TR	700-1200	500-1000
53	TL	700-1200	500-1000
	TR	1000-1500	500-1000
	BL	800-1500	
54	TL	300-800	300-800
	TR	800-1200	500-1000
	BL	300-800	300-800
	BR	800-1200	500-1000
55	TL	800-1200	500-1000
	TR	800-1500	1000-2000
	BL	1000-1500	
	BR	700-1200	500-1000
56	TL	800-1200	500-1000
	BL	1000-1500	1000-2000
	BR	800-1200	1000-1500
57	TL	400-800	500-1000
	TR	700-1200	500-1000
	BL	400-800	500-1000
59	TL	500-800	500-1000
	TR	700-1200	500-1000
	BL	800-1500	500-1000
60	CR	400-800	500-1000
61	BC	400-800	500-1000

Pg.	Pos.	Watch	Box
62	TL	800-1200	500-1000
	TR	700-1200	500-1000
63	TL	700-1200	500-1000
	TR	800-1500	500-1000
	BL	700-1200	500-1000
64	TL	700-1200	500-1000
	TR	800-1500	500-1000
	BR	700-1200	500-1000
65	TL	1000-1500	
	TR	700-1200	500-1000
	BL	700-1200	500-1000
66	TL	1000-1500	
	TR	800-1200	500-1000
	BL	500-800	500-1000
	BR	400-800	400-800
67	TL	500-800	400-800
	TR	400-800	500-1000
68	ALL	500-800	400-800
69	TL	500-800	400-800
	TR	1000-1500	1000-1500
	BL	700-1200	500-1000
70	T	300-500	700-1500
	BL	300-500	300-500
		300-500	300-500
71	TL	100-300	500-800
	TR	800-1200	500-1000
	BL	700-1200	500-1000
	BR	500-800	500-1000
72	TL	500-800	500-1000
	TR	700-1200	500-1000
	BR	700-1200	500-1000
73	TL	300-800	500-1000
	TR	100-300	500-1000
75	TR	300-500	300-500
76	BL	300-500	300-500
77	TL	300-600	
	TR	300-600	
	BL	300-600	300-800
78	TL	300-800	
	TR	300-600	300-600
	BR	300-600	300-600
79	TL	300-800	
	BL	700-1200	200-500
	TR	300-800	
80	ALL	300-600	
81	TL	300-600	
	TR	700-1200	100-200
	BL	500-900	300-800
	BR	300-600	
82	TL	700-1200	
	TR	800-1500	
	BR	300-600	
83	TL	300-600	
	TR	200-600	
	BL	300-600	
	BR	200-600	
84	BL	1000-1500	100-200
	BR	700-1200	100-200
85	TL	300-600	
	TR	300-600	
	BR	200-400	
86	TL	300-600	
	TR	300-600	200-400
	BR	300-600	
87	TL	300-600	
	TR	500-1000	
	BL	300-600	
88	TL	300-600	
	TR	500-1000	
	BR	500-1000	
89	TL	700-1200	
	TR	300-600	
	BR	400-800	
90	TL	200-400	
	TR	300-600	
	B	300-600	200-400
91	TL	300-500	
	TR	700-1200	200-500
	B	500-800	
92	TL	300-600	
	TR	200-400	
	BR	500-800	
93	TL	700-1200	100-200
	TR	500-1000	
	BL	1000-1500	100-200
94	TL	700-1200	
	TR	500-800	
	BL	500-800	
	BR	700-1200	100-200
95	T	300-600	200-400
	BL	200-400	
	BR	1000-1500	200-400
96	BL	700-1200	100-200
	BR	300-600	
97	TL	300-600	
	TR	300-600	
	BL	300-600	
	BR	300-600	
98	TL	300-600	200-400
	TR	300-600	
99	TL	500-800	
	TR	700-1200	100-200
	BL	500-800	
	BR	500-800	
100	TL	300-600	
	TR	300-600	
	BL	300-600	
	BR	600-800	
101	BL	700-1200	200-500
	BR	300-600	
102	TL	500-800	200-400
	TR	500-800	
	BL	1000-1500	
103	TL	600-1000	
	TR	500-800	
	BL	500-800	
104	TL	300-600	
	TR	200-400	
	BL	500-800	
	BR	300-600	
105	TL	500-800	
	TR	300-600	
	BR	100-200	
106	TL	700-1200	100-200
	TR	300-600	
	BR	300-600	
107	TL	300-800	
	TR	200-400	
	BL	400-700	
108	TR	700-1200	200-500
	TL	500-800	
	BR	700-1200	200-500
109	T	500-1000	
	BL	500-1000	
	BR	700-1200	
110	TL	300-600	
	TR	500-800	
	BR	300-800	
111	TL	300-600	
	TR	500-800	
	BL	500-1000	
	BR	700-1200	100-200
112	TL	500-800	
	TR	300-600	
	BR	300-600	
113	TL	700-1000	
	TR	1000-1500	
114	TL	700-1200	100-200
	TR	300-600	
115	TL	500-1000	

Pg.	Pos.	Watch	Box	Pg.	Pos.	Watch	Box	Pg.	Pos.	Watch	Box	Pg.	Pos.	Watch	Box
	TR	1000-1500		144	TL	200-400		173	TL	300-500			TR	200-400	
	BL	300-600			C	200-400			TR	100-300			BL	400-600	
116	TL	700-1200	100-200		BR	200-400			BL	200-400			BR	200-400	
	TR	300-600		145	TL	300-500			BR	300-500		262	TR	200-400	
	BR	300-600			C	300-500		174	TL	300-500			BR	100-200	
117	BL	300-600			BR	200-400			TR	300-500		263	TL	300-500	
	BR	200-400		146	TL	300-500; 300-500; 1200-1800			BR	300-600			TR	300-500	
118	TL	300-600						175	TL	200-400			BL	200-400	
	TR	300-600			TR	600-1000			TR	300-500		266	TL	100-200	
	BL	700-1200	200-500		BL	200-400			BL	200-400			BR	300-500	
	BR	300-600		147	TL	300-500		178	T	200-400		267	T	800-1200	600-1500
119	TL	300-600			TR	300-800			BL	100-200			B	800-1200	600-1500
	TR	300-600			BL	300-500			BR	300-600		268	TL	800-1200	600-1500
	BL	700-1200	100-200	148	TL	300-500		179	TL	300-600			TR	300-500	
	BR	500-800			TR	300-500			TR	200-400			BR	300-500	
120	TL	500-800			BR	200-400			BL	200-400		269	TR	300-500	
	TR	300-600		149	TL	300-800		180	TL	100-300			BR	300-500	
	BL	300-600			TR	200-400			TR	100-300		270	TL	300-500	
	BR	300-600			BR	200-400			BR	200-400			TR	200-400	
121	TL	500-1000		150	CL	300-800		181	TL	200-400		271	TL	600-1000	600-1000
	TR	500-1000			TR	600-1000			TR	300-500					
	BR	300-600			BR	400-700			BL	300-500					
122	TL	500-1000		151	TL	300-600		182	TL	300-500					
	TR	400-800			TR	300-500			C	300-600					
	BL	300-600			BL	200-400			BR	300-600					
	BR	400-800		152	TL	200-400		183	ALL	200-400					
123	TL	500-1000			TR	300-500		233	TR	300-500					
	TR	500-800			B	300-600			BL	300-500					
	BL	500-800		153	TL	300-600		234	ALL	300-500					
	BR	800-1200			TR	400-700		235	TL	300-500					
125	TR	200-400			BL	300-500			BL	300-500					
126	TL	200-400		154	TL	300-500			BR	300-500					
	TR	300-600			TR	300-500		236	TR	300-500					
	BR	300-600			B	500-900			BR	400-600					
127	TR	100-200		155	TL	400-700		237	TR	200-400					
	BL	100-200			TR	400-700			BR	300-500					
	BR	200-400			BL	300-600		238	TL	200-400					
128	TL	200-400		157	TL	300-500			C	200-400					
	TR	100-200			TR	400-600			BR	300-500					
	BL	100-200			BL	400-700		239	TL	200-400					
	BR	300-500			BR	400-600			BR	200-400					
129	TR	100-200		158	TL	300-600		240	BR	200-400					
	BL	300-500			TR	500-900		241	TR	300-500					
	BR	300-500			BR	500-900			BL	300-500					
130	BR	100-200		160	TL	500-700		242	TL	200-400					
131	ALL	100-200			TR	300-600			TR	800-1200					
132	TL	100-200			BL	500-900			BL	200-400					
	TR	300-500			BR	300-500			BR	300-500					
	BL	100-200		161	ALL	300-600		244	TL	300-500					
	BR	300-500		162	T	300-600			BR	400-700					
134	ALL	300-500			BL	500-900		245	TL	300-500					
136	TL	200-400			BR	200-400			BL	500-800					
	TR	200-400		163	TL	100-200			BR	500-800					
	BR	400-800			TR	300-500		247	TR	100-200					
138	TL	200-400			BL	300-600			BL	100-200					
	TR	200-400		164	TL	200-400		248	ALL	100-200					
	BL	100-300			BL	100-200		249	ALL	200-400					
	BR	200-400			BR	300-600		250	TL	100-200					
139	TL	300-500		165	TL	200-400			TR	200-400					
	TR	300-500			TR	200-400			BR	100-200					
	BL	400-800			BL	100-200		252	BR	100-200					
	BR	100-300		166	ALL	200-400		253	TL	100-200					
140	TL	300-500		167	TL	300-500			TR	200-400					
	BL	100-300			TR	400-700		254	TL	200-400					
	BR	200-400			BL	300-500		255	TL	100-200					
141	TL	500-800		168	TL	200-400			TR	200-400					
	TR	300-500			TR	300-600		257	TL	200-400					
	BL	100-300			B	400-700			BR	200-400					
	BR	100-300		169	TL	300-600		258	TL	300-500					
142	TL	300-600			TR	200-400			TR	300-500					
	TR	200-400			BL	100-300			BR	200-400					
	BR	500-800			BR	100-300		259	ALL	200-400					
143	T	300-500		171	ALL	100-300		260	TL	300-500					
	BL	200-400		172	ALL	100-300			BR	300-500					
	BR	200-400						261	TL	200-400					